Basic income is one of the most innovative, powerful, straightforward, and controversial proposals for addressing poverty and growing inequalities. A Basic Income Guarantee (BIG) is designed to be an unconditional, government-insured guarantee that all citizens will have enough income to meet their basic needs. The concept of basic, or guaranteed, income is a form of social provision and this series examines the arguments for and against it from an interdisciplinary perspective with special focus on the economic and social factors. By systematically connecting abstract philosophical debates over competing principles of BIG to the empirical analysis of concrete policy proposals, this series contributes to the fields of economics, politics, social policy, and philosophy and establishes a theoretical framework for interdisciplinary research. It will bring together international and national scholars and activists to provide a comparative look at the main efforts to date to pass unconditional BIG legislation across regions of the globe and will identify commonalities and differences across countries drawing lessons for advancing social policies in general and BIG policies in particular.

Series Editors:

Karl Widerquist is an associate professor at SFS-Qatar, Georgetown University.

James Bryan is an associate professor of Economics at Manhattanville College.

Michael A. Lewis is an associate professor at Hunter College School of Social Work.

Basic Income Reconsidered
Simon Birnbaum

Alaska's Permanent Fund Dividend
Edited by Karl Widerquist and Michael W. Howard

Basic Income Guarantee
Allan Sheahen

Basic Income Guarantee and Politics
Edited by Richard K. Caputo

Exporting the Alaska Model
Edited by Karl Widerquist and Michael W. Howard

Citizen's Income and Welfare Regimes in Latin America
Edited by Rubén Lo Vuolo

Independence, Propertylessness, and Basic Income: A Theory of Freedom as the Power to Say No
KarlW iderquist

Basic Income and the Free Market: Austrian Economics and the Potential for Efficient Redistribution
Edited by Guinevere Liberty Nell

Basic Income in Japan: Prospects for a Radical Idea in a Transforming Welfare State
Edited by Yannick Vanderborght and Toru Yamamori

Free Money for All: A Basic Income Guarantee Solution for the Twenty-First Century
MarkW alker

Free Money for All

A Basic Income Guarantee Solution for the Twenty-First Century

Mark Walker

palgrave
macmillan

FREE MONEY FOR ALL

First published 2016 by
PALGRAVE MACMILLAN

Palgrave Macmillan in the UK is an imprint of Macmillan Publishers Limited, registered in England, company number 785998, of Houndmills, Basingstoke, Hampshire, RG21 6XS.

Palgrave Macmillan in the US is a division of Nature America, Inc., One New York Plaza, Suite 4500, New York, NY 10004-1562.

Palgrave Macmillan is the global academic imprint of the above companies and has companies and representatives throughout the world.

Hardback ISBN: 978–1–137–47132–1
E-PUB ISBN: 978–1–137–47134–5
E-PDF ISBN: 978–1–137–47133–8
DOI: 10.1057/9781137471338

Distribution in the UK, Europe and the rest of the world is by Palgrave Macmillan®, a division of Macmillan Publishers Limited, registered in England, company number 785998, of Houndmills, Basingstoke, Hampshire RG21 6XS.

Library of Congress Cataloging-in-Publication Data

Walker, Mark, 1963–
Free money for all : a basic income guarantee solution for the twenty-first century / Mark Walker.
pages cm.—(Exploring the basic income guarantee)
Includes bibliographical references and index.
ISBN 978–1–137–47132–1 (hardback : alk. paper)
1. Guaranteed annual income. 2. Income maintenance programs.
I. Title.

HC79.I5W334 2015
362.5′82—dc23 2015021526

A catalogue record for the book is available from the British Library.

Contents

Figures and Tables

Figures

Tables

Acknowledgments

I am grateful to a very large group of my students, without whom this book would never have been written. I teach introduction to philosophy to a large section of about 90 students each semester, the proverbial "philosophy 101" class. The vast majority are there simply to fulfill a general education requirement. Most come to the first class with little knowledge of philosophy and even less interest. I do not say this with any condemnation: a lack of knowledge and interest in philosophy is to be expected, given our current educational system. Philosophy is not a part of the mandatory high school curriculum and very few high schools offer it as an elective. As a result, I am tasked with teaching about 90 philosophy novices each semester.

I use a time-honored means to engage my philosophy newbies: I spend much of the semester arguing against some of their cherished beliefs. On the first day of class, I do a little informal poll: How many students believe in god? How many students believe in capitalism? And so on. Given the demographics of my classes, it is no surprise that more than 90 percent believe in god and capitalism. I then tell them that I will spend the semester arguing against these beliefs. Previously, I used to spend a few weeks arguing for socialism. About five years ago I decided my course needed a little refreshing, so I started arguing in favor of a basic income guarantee (BIG)—a monthly payment from the government, sufficient to eke out an existence—rather than for socialism.

Surprisingly, I found students initially were much more hostile to BIG than to socialism. I say this is surprising because switching from a capitalist to a socialist economy is a far more radical change than choosing BIG. I suspect the difference lies in the fact that talk of socialism tends to seem very hypothetical to students, whereas they can see just how easily BIG could be implemented. In other words, BIG seems "more real" than socialism, which means it is a bigger threat to the social reality that students are accustomed to.

Although they are novices, I am always impressed by how quickly some students take to philosophy. The class discussions I have with students tend to be a perfect microcosm of the public and academic debates about BIG. The most common objection from students, which perfectly parallels the public and academic debates, is this: the idea is crazy because some people would simply collect the guaranteed income and not work. *Free Money for All* is a sustained attempt to show that the idea is not crazy, and we should hope that some will not want to work. The arguments offered here were hammered out on the forge of my philosophy 101 classes over the past five years. I thank my students for their patience and for their spirited participation in our discussion of economic justice. I hope my former students will be amused by the thought that when we adopt BIG as policy and it becomes the new status quo, I will have to argue against BIG.

Students in my upper division social and political philosophy class have also been involved with the arguments of this work. In particular, I would like to thank Henry Allingham, Angela Bischoff, Zane Bischoff, Jamie Bronstein, Brian Kraus, Henry Allingham, Michael Heinz, and Shelbi Meisner. Their input has been invaluable. There are probably other student names I have forgotten. My only defense is that it is always easier to remember the troublemakers.

I organized a conference on BIG in 2014, the Little BIG Conference, which was hosted by my department. Funding was provided through the Richard L. Hedden Endowment to the philosophy department at New Mexico State University. I would like to thank Richard Hedden for making the conference possible, and also for that small matter of my job, which was made possible through the same endowment. I learned a lot at this conference and I would like to thank each of the participants: Vaughn Bryan Baltzly, Juliana Bidadanure, Jamie Bronstein, Jurgen De Wispelaere, Mike LaTorra, Attila Mraz, and Karl Widerquist. My good friend James Hughes was also invited, but the weather thwarted his travel plans. James and I look forward to the day when the weather is controlled through technology.

I have some wonderful colleagues in the philosophy department. I would like to thank Lori Keleher, Danny Scoccia, and Jean-Paul Vessel for allowing me to test-drive some of the arguments with them. I would also like to thank (yet again) Jamie Bronstein. She is a historian by training but, I suspect, a philosopher by inclination. Jamie has patiently listened to many of the ideas in this book and offered detailed written comments on most of the work.

I am not sure about the etiquette of having favorite students. I will confess that in the thousands of philosophy 101 students I taught over the years, three deserve special mention. My daughters, Chantal

Walker and Danielle Walker, have both taken my philosophy 101 class. It was a real pleasure to have them in class to see what I do for a living. I am pleased to say too that it did not take much to persuade them about the viability of BIG. I assume this is because they were highly attuned to respond to excellent arguments—either that or the Stockholm syndrome had set in.

My wife, Dawn Rafferty, is a philosophy 101 survivor as well. Dawn has helped me with this project at every stage. Often when I get philosophical ideas, we talk about them while we walk Coco in the mornings. And by saying "we talk about them," I mean I talk and she suffers. Dawn has also edited three different versions of this book. It is with love and gratitude that I dedicate this book to her.

CHAPTER 1

Basic Income Guarantee

Introduction

Imagine this scenario. You are an anthropologist leading a team to investigate reports that there are human fossils in Antarctica. On the way there, to avoid a menacing cyclone, the ship's captain deviates from the planned route into less-traveled waters. The detour is lucky for you. You stumble onto the find of a lifetime: an unknown population on a remote South Pacific island. After checking several databases just to be sure, you confirm that the premodern indigenous population has almost certainly had no contact with the rest of the world for centuries, perhaps millennia. Although professional and international treaties prohibit you from brazenly making first contact, your team consults with UN officials via radio and is given permission to study the tribe from afar—so long as you remain undetected.

Each member of your team is assigned to observe several individuals of the tribe with binoculars and telescopes. You are tasked with observing four males. Peering through your telescope you quickly notice that two appear to be well fed and enjoy elaborate shelters. Their dwellings seem to provide ample protection from the remnants of the cyclone that is still pelting rain. Your other two subjects appear to be malnourished. At night, they sleep exposed to the rain and the wind. As far as you can tell, they have no shelter to call their own. You are astonished. You are well aware that there is anthropological evidence of some material inequality among members of premodern tribes; still, you are astounded by the extreme level of inequality in this tribe.[1] It is not as if the tribe as a whole is short on food or shelter. Indeed, the tribe actually has a vast surplus of food, and much seems to get wasted. Furthermore, your team confirms that there are enough shelters such that everyone could be protected. Your two materially well-off subjects actually appear to own several homes, meaning that some huts remain unused every night. Your team tries to puzzle this out. Why not simply let those exposed to the elements (including your two subjects) use the huts that no one else is using? As professional anthropologists, your group was taught long ago that the term "primitive culture" is colonialist and ethnocentric. Still, this and similar terms slip out during your team discussion as you ponder some hard questions: What conventions of the distribution of material goods could justify such remarkable differences in access to material necessity? how can some be forced to sleep outside in the elements, short on food, while others live in material opulence? It is both puzzling and exciting that this premodern tribe differs from so many others studied by anthropologists. Good fortune has smiled upon you; your career is made.

You return to the United States to much fanfare. You do a number of interviews about your remarkable discovery of the premodern tribe and do the late night talk show circuit. All the attention is exciting, but you are glad when the hubbub dies down. You return with relief to your hometown, Anywhere, to resume your teaching and research.

You and your spouse plan a nice dinner out to celebrate the return to normalcy. You pull up to the restaurant just in time to witness this scene. Two very expensive cars, known for their fine German engineering and worth three times the average yearly US salary, pull up just down the street from you. Their well-dressed owners exchange some good-natured ribbing about the shortcomings of each other's vehicle as they enter the most expensive restaurant in town. On the same street, two homeless persons beg for change. One holds up a sign indicating that he is a US veteran down on his luck. You wonder if it is colonialist and ethnocentric to describe your own culture as "primitive."

Of course, this thought experiment is a bit forced and unrealistic. As is known, premodern societies typically never approach anywhere near the material inequality suggested here. Such a discovery would really be remarkable. On the other hand, that part of the thought experiment where two homeless persons share a public street with two owners of fine German engineering is not the least bit unrealistic. We have all seen homeless people living on our streets; we have all also seen pictures of the rich with their private planes, private yachts, private estates, and private country clubs. The question discussed in this work is whether such disparities can be justified.

From the point of view of capitalism, the scene from Main St., Anywhere Town, is a triumph of distributive justice. The capitalist market rewards the talented and hardworking while it fails to reward the untalented and lazy. The capitalist vantage says that the owners of the expensive cars almost certainly must be hardworking and talented; the homeless must be untalented or lazy, or perhaps both. And if it is just a matter of bad luck, well, some are lucky and some are not.

Yet for many of us, the scene is incredible. How can all these four persons on this stretch of public road be citizens of one and the same community? Perhaps some differences in wealth are justified, but what could possibly justify such an extreme inconsistency of wealth distribution? From our vantage point, the scene seems almost surreal. But what is to be done about it? Socialism, many think, has failed. So it seems we are left with capitalism.

A Middle Course?

A policy that looks to steer a possible middle road between socialism and capitalism stems from an idea that has waxed and waned for some time: a Basic Income Guarantee (or BIG for short). The name explains much. The idea is to guarantee every citizen in the country an unconditional income sufficient to meet some minimal threshold. BIG and related proposals have a long history often traced back at least to Thomas Paine's *Agrarian Justice* (1796), sometimes as far back as the Bible.[2]

On the face of it, BIG sounds like a compromise. It would not be anywhere near the level of economic equality sought by many egalitarians and socialists, yet it would do much to stem the worst excesses of unrestrained capitalism. It would, for example, help the economic situation of the homeless veteran and his downtrodden compatriot in our example.

Although much of the argument applies also to other developed nations, this work deals primarily with the United States and argues for an annual BIG of $10,000. In other words, the aim is to support the conclusion that every US citizen should receive $10,000 with no conditions attached. It is unconditional in the sense that both rich and poor alike should receive it, with no work requirement. So, unlike the current welfare policy, BIG would not be conditional on being employable or looking for work. The perennial Malibu surfer, who has never worked a day in his life, would be as much entitled to his BIG as the 80-hours-a-week workaholic.[3]

Why should the perennial surfer be entitled to a guaranteed income? I admit this is a good question. Before outlining the arguments in favor, it will help us to first get a better idea of exactly what BIG is.

The proposed annual BIG of $10,000 would be paid directly by the Internal Revenue Service proportionately once a month. So, each month, every US citizen would receive one-twelfth of his or her $10,000 per annum, or $833 per month, payment as a direct deposit into his or her account. As indicated, this same amount would go into the accounts of the richest citizens, including Bill Gates and Warren Buffett, as well as our surfer.

Most discussions of BIG shy away from providing a precise dollar amount. I can certainly see the reasons for this reticence. Often the point is to discuss the theoretical pros and cons of BIG and not get bogged down by the details. I have decided (perhaps unwisely) to go against this trend.[4] My thought in doing so is to prove that BIG is perfectly feasible. In other words, the intention is to show that BIG is not some utopian dream but rather a policy we could easily adopt. The danger is that in putting a hard number out there, I risk alienating some of those who might otherwise be friendly to the suggestion. Some may wish for a larger BIG, others for a smaller sum.

It would be a mistake to think that the word "basic" in BIG will settle the issue of an appropriate amount. One potential problem in stating an amount stems from the vagueness and ambiguity of the word basic. What exactly does it mean? In one sense, it might mean the bare minimum to avoid dying of starvation or exposure. In that sense, perhaps $10,000 per annum is far too much. Nine people might be able to share the rent on a bachelor apartment, perhaps three to a bed, sleeping in shifts. Eating noodles and peanut butter sandwiches for many meals might help keep the food bill down. Perhaps by living right on the edge like this, someone might survive on $5,000 a year in parts of the United States.

On the other hand, one might think that basic means something like "minimally decent." On this understanding, an income must be sufficient to live on one's own, or at least not require bunking with eight others in a bachelor apartment, and still have enough for a cell phone, Internet, TV, a clothing allowance, and so on. Perhaps the income level in that case might need to be at least $15,000 per year. Even with these two senses of basic, there is a certain amount of vagueness: does the having "enough to eat" mean eating until one is full or having just enough calories to survive? And on the more expansive understanding, we might wonder whether Internet access means high-speed cable or a slower dial-up connection. Does TV access include some of the cable packages with educational channels? And so on.

Invoking the "poverty line" does not help much either.[5] The same problems reappear. What counts as "poverty"? Surely, starving for lack of food does, but should people have enough for a high-speed Internet connection, and so on? Even the US federal government cannot decide on a poverty line. They offer two different numbers: poverty threshold and poverty guidelines.[6] Both are between $11,000 and $12,000 per year per person. So, someone living on BIG as their sole source of income would live below the poverty threshold or guideline. Still, this does not give us much idea about whether BIG is enough without knowing the details of what is involved in calculating the poverty threshold or guideline.

More helpful is the living wage calculator on the Massachusetts Institute of Technology website as one of their "Poverty in America" resources. In part, the living wage calculator on the website is designed to show the minimal costs to live in selected areas. Below are data for two different locations: Las Cruces, New Mexico and New York (table 1.1).

The first thing to notice is that the living wage calculator puts a $10,000 BIG as well below the living wage in both places. The lion's share of the difference between the Las Cruces and New York costs is on housing, $479 versus $1,129, respectively. We will talk more about this difference in the next section, but for the moment, we will concentrate on the Las Cruces numbers. BIG is not taxed, so we can see immediately that the appropriate comparison is for the after-tax amount. This still leaves a deficit of $4,544 in terms of annual income, or $379 per month. (The living wage calculator comes to $1,212 per month while a $10,000 annual BIG payment works out to $833 per month.) One thing that could be pared down from the

Table 1.1 Livingw agec alculator[7]

	Las Cruces	*New York*
Food	$242	$242
Child care	$0	$0
Medical	$137	$136
Housing	$479	$1,129
Transportation	$285	$262
Other	$69	$104
Required monthly income after taxes	$1,212	$1,873
Required annual income after taxes	$14,544	$22,476
Annual taxes	$2,355	$4,045
Required annual income before taxes	$16,899	$26,521

living wage calculator is transportation. A bus pass in Las Cruces costs $30 a month, which would shave $255 off a month on account of transportation, bringing the gap to $124 a month. I am not saying that the living wage calculator is wrong in having transportation at that amount. Their calculation is based on the idea that someone is working for a living. For many jobs, it would be impossible to take a bus to work in time. Since we are working on the idea that someone is trying to live off BIG as the sole source of income, the cost of transportation to work is not an issue to be considered. One of the sacrifices to make would be the time and convenience associated with faster modes of transportation, such as a personal vehicle or taxi rides. Another savings under the present proposal is a universal health care system (discussed in chapter 2) that would eliminate most of the $137 a month in medical bills. Further savings can come from shared accommodation. It is possible to rent a three-bedroom place in Las Cruces starting at $500 a month. Utilities (gas, electricity, water, and Internet) might work out to somewhere around $150 to $320 a month, resulting in a savings from the living wage budget of $159 a month, more than enough to cover the aforementioned gap.

Here is another example. While consulting on this issue, I have turned to experts. I have asked students in my university classes how much they live on. I do not intend to be facetious, but I live in southern New Mexico, one of the most impoverished areas in the United States. Many students at my university are first-generation university students from lower income families. Here is what I have found. At New Mexico State University, you can live in a triple-occupancy dorm for $1,196 per semester and have unlimited food for $1,673 per semester. At a per-annum rate, this works out to $8,607. Triple occupancy in a small room is pretty harsh, I admit. However, it includes cable/Internet and unlimited hot water. The food bill includes restaurant-type meals for as many times a day as one wishes. The meals are from a large cafeteria/smorgasbord type of setup. The reason many students in my classes do not live in the dorms is because it is still too expensive for them. Students find off-campus housing in shared houses/apartments that give them more space, and the cooking facilities reduce their expenses. My point is that many college students manage to eke out a living on $10,000 a year. Indeed some people, including some of my students, would find this amount to be positively opulent compared to what they are forced to live on at the moment.

In wading through some of these numbers, I have tried to show that there is not much point in arguing about the meaning of terms like "basic income" or "poverty line" in the abstract. They are too elastic to resolve disputes. It is not that they are without meaning: $2,000 a year is below any reasonable threshold for poverty or basic income; $30,000 is well above. In arguing about where the line should fall between these extremes, we have to make specific arguments about what sorts of things one should be able to buy on a basic income to be above the poverty line. I have indicated the sort of extremely frugal lifestyle that living on BIG alone would impose—a perennially impoverished college lifestyle. I would like to see it set higher, but I will not argue for that here. If I were king of the United States, I would set it between $15,000 and $20,000 and perhaps adjust it (up or down) according to societal changes. The higher the number, the harder it is to argue for, and the less likely it will be accepted. Any number has to be a compromise, so I admit $10,000 as a compromise. Still, it is a compromise worth fighting for. As I will indicate below, the number could be tweaked in response to societal changes. Also, there are good reasons to believe that with advances in technology the purchasing power of $10,000 will increase.

Unconditional Experiments in Living

I have talked about BIG in a number of my classes. Although I have now come to expect it, initially I was quite surprised by the adverse reactions of students to the very idea of BIG. When broaching this subject with students, I think I would get a warmer reception if I were to advocate a policy of eating human babies. This extreme aversion to BIG may explain why, when I ask my students what they would do if they lived in a society that offered a BIG of $10,000, their answers are so ill conceived, at least at first. Usually the discussion starts with a student stating that he (it is always a "he") would drink beer every day and play video games. Classmates quickly point out that BIG would not provide sufficient income to drink the proposed 12 beers per evening. Others pipe up, and this would soon, very soon, get tiresome.

Perhaps somewhat surprisingly, most students, on reflection, say that it would not change their plans very much at all. They say that BIG is hardly sufficient for the sort of material lifestyle they hope for, and so they will still need to go to school and get a job. They do say that it would make things easier. For example, it would greatly reduce the need for student loans while going to school. A certain number mention that it would open up the possibility of going to graduate school instead of trying to find work to pay off their student loans.

Some students mention starting their own business. It is easy to see how BIG might support increased entrepreneurial activity. One of the major reasons so many small businesses fail in the first couple of years is because of lack of capital. Often this is the result of having to pay the living expenses of at least one person: the entrepreneur. BIG might allow more would-be entrepreneurs to avoid this pitfall. Consider that there are a number of small businesses, for example, Internet companies or home repair businesses, that can be started for very little capital. A computer and an Internet connection are minimal requirements in the former case, and tools and a truck, in the latter. The biggest need for capital in both cases is typically the expense of covering living expenses while the company takes wing.

Certain artistic lifestyles would also be pursued with BIG. Some of my students have expressed interest in pursuing their dreams of being in a band fulltime. Most musical careers do not pay particularly well. With BIG as a backstop, more might take the risk and follow their dreams rather than resign themselves to a soul-crushing job merely for financial security. Others have expressed interest in pursuing poetry fulltime or selling wood carvings on the Internet. Many of these dreams would not be reasonable options without BIG, as the threat of financial destitution would be too great.

Another possibility would be to live a relatively simple life as advocated by Thoreau. Rural property is relatively cheap. It is possible to buy an acre for less than $1,000. It would be possible to build a house, plant a garden, install solar and wind generation of power, and live "off the grid" on BIG. A couple devoted to such a project might soon be able to bank some of their BIG money, as their monthly expenses would be quite low.

BIG would enable many to be socially productive in a manner that is not often recognized by traditional market-driven values of distribution, for example, withdrawing for a time from the workforce to look after an ill parent, or to homeschool children. These and other "caregiving" relations help society in the aggregate but are usually not recognized as monetarily valuable activities.

Indeed, it would also be possible to spend all one's time surfing, so long as one were content to live on a mere $833 a month. Rawls's Malibu surfer would have some difficulty though, given the high cost of housing in Malibu. But in theory, it would be possible to surf all the time where rent is cheap and waves are plentiful. My students who think they would be content playing video games for the rest of their lives would also have the means to do so with BIG.

If they want to drink beer every evening as well, they may have to make their own. Still, if they are willing to live frugally, it might be possible.

Naturally, it will be surprising to know how many people will work if they have access to BIG. Many object to BIG on precisely these grounds: BIG will result in collapse of the economy as not enough people will be willing to work to keep society going. We will postpone discussion of this objection until chapter 8. The tentative reply is that our biggest worry about the future should not be about too few people looking for work, but too many.

One thing we should underscore is the absolutely unconditional nature of BIG. Even though he is worth billions upon billions, Warren Buffett would be as entitled to BIG as a homeless veteran. BIG is also unconditional in terms of work choices: BIG would be the same for those not working at all, part-time workers, and people working two or three jobs at a time.

BIG as Policy: Capitalism, Socialism, and the Welfare State

I mentioned above that BIG is sometimes seen as a compromise between capitalism and socialism. It will help to explore this suggestion.

The first thing to understand is that BIG is a policy recommendation; BIG is not a comprehensive theory of economic justice of the sort offered, for example, by proponents of capitalism and socialism. It might help to illustrate the difference between policy and comprehensive theory with an example from criminal justice. Suppose a government is considering a change in policy with respect to capital punishment. Presently, capital punishment is used with some regularity, but there is support for change. Different, but incompatible, views of criminal justice might agree that there should be a policy change. We can imagine support coming from the following members of the state legislature:

1. Pacifists who believe the taking of life is always wrong.
2. Retributionists who believe capital punishment is justified in theory, but that too many innocent people—particularly racial minorities—have been put to death. These members support the banning of capital punishment as a reasonable measure until this problem can be solved.

3. Deterrentists who believe that capital punishment is justified only if it can be shown that it effectively deters more crimes than life in prison. Since empirical studies have shown no such deterrence effect, these members are for a change in policy.
4. Members who are concerned simply with budget issues: they believe it is more cost efficient to have prisoners serve life in prison than pay for the long and protracted legal battles by the condemned.

The important point here is that members agree on *what* is to be done—capital punishment is to be prohibited—but they disagree on *why* it is to be done. They agree on the policy question, but their philosophical reasons for doing so differ.

When speaking to the people about BIG, I often get the suggestion that BIG is not even a compromise between capitalism and socialism, it is a policy that only a socialist, or at least someone far along the left wing of the economic spectrum, might endorse. However, famously, the Noble winning proponent of capitalism, economist Milton Friedman, endorsed a proposal similar to BIG. A negative income tax was his solution to eliminating poverty.[8] Of course, not all theorists who endorse capitalism would support BIG. Indeed, there is a tradition in American political theory and activism that suggests that (most) taxes are theft and slavery.[9] So it is plausible to assume that many apologists for capitalism would adamantly reject BIG.

There is a long history of socialist thinkers endorsing BIG. In the early twentieth century, Bertrand Russell offered BIG as a means to greater freedom.[10] Recently, the Marxist socialist Erik Olin Wright has suggested BIG as a "real utopia," a feasible means to help the most materially disadvantaged in society.[11] While some socialists have endorsed BIG, it is certainly not true that all socialists would endorse it. After all, even though BIG would reduce some of the worst forms of poverty, it would hardly result in the ideal egalitarian society envisioned by many socialists. BIG is consistent with, for example, several people having to share a bachelor apartment while others own palatial mansions—something that many socialists might find unjust even if it is seen as an improvement over the status quo.[12]

If there is any legitimacy in the idea that BIG as a policy strikes a compromise between capitalism and socialism, then it would seem that BIG would find a natural home among BIG "welfare liberals"—apologists for the welfare state. The ideal of the modern welfare state is sometimes promulgated along these lines. The welfare state attempts to take what is best in capitalism and socialism and weld it into a workable compromise. It permits a certain amount of capitalism while the state reaps some of the economic benefits (in the form of taxes) of capitalism to ensure the least materially advantaged in society do not die from material want. Members of the so-called G7—the United States, the UK, France, Germany, Italy, Canada, and Japan—have been termed "welfare states" in this sense. This is called an "ideal" since it is not clear whether all the G7 members actually meet the criteria. For example, US citizens by the tens of thousands die each year from lack of access to basic medical care.[13] As noted above, there are people starving in the streets of the United States, digging out garbage to try to survive. So the reality is that the ideal of the welfare state is far from realized in the United States. Conditions in the other G7 nations tend not to be as bad for the poor, but most could do much better according to the percepts of many welfare liberals.

Although the welfare state itself is sometimes advertised as a compromise between capitalism and socialism, important questions remain about the compatibility of BIG and the welfare state (even the unrealized ideal of the welfare state). First, for the most part, in *practice*, the welfare state is understood as having a "willingness to work" requirement for grant of welfare; specifically, one has to prove that one is unable, for physical or mental reasons, to be able to work, or that one is looking for work.[14] In other words, the welfare state offers welfare conditionally, not unconditionally. For example, under the present system in the United States, if one is laid off, then one may be entitled to unemployment insurance after a certain number of weeks. When unemployment insurance runs out, then one may be entitled to welfare after one proves to some government official that one is actively seeking work.

Turning from practice to theory, welfare liberals are divided on the question of BIG. The most prominent welfare liberal is John Rawls. In his magisterial *A Theory of Justice* and other works, Rawls provides a defense of many aspects of the current welfare state.[15] Like Marx, Rawls sees that capitalism has helped increase the material prosperity of society as a whole. However, according to Rawls, a laissez-faire capitalism is not justified. Rather, a conditional form of capitalism is justified. Private ownership and the profit motive are justified to the extent that they improve the material well-being of the most economically disadvantaged. This is Rawls's famous "difference principle": "social and economic inequalities are to be arranged so that they are both to the greatest benefit of the least advantaged."[16] We will postpone examining whether Rawls is justified on this point. For the moment, we just need to understand it. At least in outline, the idea is simple enough. Imagine if no economic inequalities were permitted. According to Rawls, under such a system of economic distribution, everyone would be worse off. In this model, the work incentive would be lost. The most talented and the most productive would have no reason to work hard. This means that there would be less production of material goods for everyone, including those unable to produce. So, says Rawls, we should organize society such that there is sufficient motivation for the most productive and talented to work hard. Such a system would permit some to own more than others, but some of the production of the most talented and productive would go to the least advantaged. In practice, these resources would be distributed to the least well-off through taxes. One might describe this as redistribution of income from the rich to the poor, but this would be misleading. Rawls's theory of justice permits large material inequalities only to the extent that they help the worst off. So a rich person who claims that taxes are theft is wrong, at least according to Rawls. Any income a rich person generates is not really the rich person's until the least well-off are made as well-off as possible. (We will explore this subtle point further below.)

It would seem then that Rawls's welfare liberalism and BIG are a natural fit. After all, BIG looks like an efficient way to satisfy the difference principle, and BIG looks to do more for the least advantaged than current welfare policies. However, it seems that Rawls is, at best, ambivalent on the question of the fit between BIG and his theory of justice. In *A Theory of Justice*, Rawls mentions in passing the possibility of something like BIG (which he refers to as a "negative income tax").[17] In a later and, I believe, more considered treatment of the issue, Rawls writes:

> Those who are unwilling to work would have a standard working day of extra leisure, and this extra leisure would be itself stipulated as equivalent to the index of primary goods of the least advantaged. So those who surf all day off Malibu must find some way to support themselves and would not be entitled to public funds.[18]

To understand what Rawls claims here, we need to understand what he means by "primary goods." Primary goods are "things that every rational man is presumed to want."[19] Examples from *A Theory of Justice* include intelligence, imagination, health, civil and political rights, liberty, income and wealth, and the social bases of self-respect. In his later work, Rawls adds leisure time to the list of primary goods. His reason for denying the Malibu surfer BIG is that the publicly supported surfer would then have more of some of the primary goods than others. In other words, according to the later arguments of Rawls, the Malibu surfer who demands public support would in effect be asking for more of the primary goods because he is asking for more leisure. This would be as unjust as persons asking for more liberty or civil or political rights than their peers. Rawls's position on BIG and the Malibu surfer has been criticized even by those who are generally sympathetic to his theory of justice.[20]

The reason I have mentioned this disagreement is to underscore the fact that the use of "welfare liberal" and "welfare state" in this work is more specific than the way these terms are sometimes used. I will use welfare liberal to indicate proponents who believe, as part of their ideal normative theory, that welfare should be conditional on some sort of willingness to work requirement; and welfare state as one where some sort of work (or willingness to work) requirement should be a condition for receiving public funds. I am not suggesting that this is the correct description of welfare state or welfare liberal, only that it will help us to keep track of different positions.

Note too that there is nothing inherently contradictory about a proponent of the welfare state endorsing BIG. This is because there is always a gap between one's normative theory and its implementation into policy. Recall our earlier example where the retributionist supported banning capital punishment due to imperfections in the system. The argument was that although capital punishment is justified when the criminal justice system is working properly, when innocents and a disproportionate number of racial minorities are executed, capital punishment ought to be suspended until the problem is resolved. Similarly, a welfare liberal might say that in times of high unemployment, BIG might be a reasonable solution. After all, the idea that people ought to be looking for work seems to assume that there will be work to be found. If jobs cannot be had, then there is no sense in claiming that people should get jobs or look for jobs. And if the number of unemployed swells, it might be much more efficient to simply provide BIG until unemployment rates return to some reasonable level than to hire a small army of government officials to make sure the multitude of unemployed are looking for jobs that do not exist.[21]

Overview of the Argument

It may help to give a preview of some of the arguments for BIG by situating it in relation to the "big three" alternatives—capitalism, socialism, and the welfare state—as well as placing the argument within the existing literature on BIG.

In general, the philosophical literature on BIG has emphasized the inadequacy of the big three. Unsurprisingly, much of the philosophical critique has to do with a work requirement. As we have seen, the welfare state makes state assistance for the poor conditional on some sort of "actively looking for work" requirement. The work requirement is not indigenous to the welfare state. Socialist states, like the former USSR, went so far as to have laws against not working, backed by prison time for those convicted.[22] Capitalism does not make work a formal requirement, but it is a de facto requirement for all but the rich. What this means for freedom, then, is that the big three do not provide freedom with respect to the question of work. For example, Karl Widerquist argues that any sound political philosophy must recognize the freedom of citizens to opt out of distributive arrangements of society at large.[23] If one lives in a generally capitalist society, then BIG must be provided so that citizens have the ability to opt out of a capitalist distributive arrangement. If one lives in a generally socialist society, then BIG must be provided so that citizens have the ability to opt out of socialist distributive arrangements. As Widerquist aptly puts it, citizens must have the "power to say no" to the economic system that they find themselves in.

So, one theme we find in the existing literature is a critique of the concept of freedom. According to this critique, freedom is understood too narrowly by the big three. It must be expanded, and a more adequate understanding of freedom requires BIG. [24]

A second important theme in much of the current literature on BIG is that freedom is to be conceived of as a right. This work sails against the current of much thinking about BIG by framing it in terms of good consequences rather than rights. We will have to postpone a fuller discussion of rights versus good consequences until chapters 4 and 8, but briefly, we may illustrate the dispute with the following thought. Consider the question of whether to let someone continue to spout racist remarks on a busy street corner. The corner in question is unique in that it has traditionally attracted many tourists because of its historical significance; however, tourists are now avoiding the area because of the offensive speaker. Local businesses are suffering and have to lay off workers because of the reduced tourist traffic. Here we might feel ourselves torn by the requirement of honoring the right to free speech and thinking about the bad consequences for the larger community when we allow the racist to negatively affect so many.

One of the main arguments of this book is that adopting BIG as public policy will lead to good social consequences. In particular, BIG will lead to increased total aggregate happiness and freedom.

The argument, then, has some overlap with the main current in BIG thinking by making a connection between BIG and freedom. But the present argument turns on the point of increasing aggregate freedom rather than honoring the right of individuals' freedom. The connection between BIG and happiness is also a departure from much of the mainstream of philosophical justification for BIG.[25]

While I have learned much from other philosophers' writings on BIG, and probably most from Philippe Van Parijs and Karl Widerquist, I have not spent a lot of time in this book trying to sort out where we agree and where we disagree. (That would have to be a much longer and a rather different book.) Rather, my main opponent is our present capitalist system. Interestingly, this opponent is, for the most part, faceless. By this, I mean that it is difficult to find political philosophers who think that our present system is just. As we noted above, Rawls endorses certain aspects of capitalism as an efficient means to generate wealth, which is to be redistributed in accordance to the difference principle. Yet, Rawls is not an apologist for contemporary capitalism in the United States, since it nowhere near approximates what is required by the difference principle.[26] So Rawlsians tend to believe that governments should do more to tame the excesses of capitalism when it comes to the concentration of wealth in the hands of a few. On the other hand, libertarian Nozick tends to think that governments go too far in redistributing wealth.

The general strategy in this book is to use an internal critique of contemporary capitalism. Given that contemporary capitalism seems to enjoy so much support, evidenced at least in part by the fact that there is little support among the electorate for political candidates who are close to Rawlsian liberalism or libertarian candidates, it is worth seeing what follows a commitment to contemporary capitalism. The argument is that contemporary capitalism contains within it the seeds of its own destruction.

One argument addresses the thought that BIG requires us to redistribute wealth in a way that is inconsistent with the capitalists' ethos. In chapter 3, I will argue that BIG and capitalism are not in tension. BIG flows as a consequence of people trying to maximize their profit on their ownership of capital. This admittedly counterintuitive position turns on the idea that citizens own an equal share in state capital, and individual shares of state capital are sufficient to generate enough return to finance BIG.

A second internal critique involves the relationship between capitalism and the consequentialism that animates this work, which we discuss in chapter 4. In particular, the concern is that maximizing good social consequences conflicts with the economic rights of individuals that many capitalists champion. A limited conclusion is reached: competitive markets require restraining the economic activities of individuals in the name of the social good of competitive markets. Since competitive markets are a constituent part of capitalism, capitalism requires restraint of individual economic activity for the benefit of society at large. In other words, it is false that consequentialism and capitalism are in conflict over the question of pursuing the social good, because capitalism itself must appeal to some such principle for its own defense. This sets the stage for later chapters where it is argued that BIG is the best means to maximize the social good. In effect, then, the chapter attempts to silence the objection by apologists for capitalism that we should not determine economic policy by overall economic good.

Chapters 5, 6, and 7 examine how BIG will serve the common goods of peace, happiness, and freedom. Chapter 5 looks at how the revolution in robotics and advanced computers threatens the aim of full employment. It is argued that BIG is a good and necessary means to keep the peace: displaced workers will be angry, and justifiably so, if they do not reap at least some of the rewards of this economic revolution.

Chapter 6 argues that BIG is the best policy in the economic realm to increase gross national happiness. The basic thought in this chapter is that a given dollar buys more happiness for the poor than the wealthy. Giving $1,000 to a homeless person will do more to boost his or her happiness than giving the same $1,000 to uber-rich people such as Warren Buffett or Bill Gates. The present distribution of wealth and income in the United States is not the most efficient for converting dollars to happiness. BIG is a better policy option for turning wealth and income into happiness.

Chapter 7 argues that BIG is the best policy in the economic realm to increase gross national freedom. The basic thought is the same as in the previous chapter. BIG is a better policy option for turning wealth into freedom.

Chapter 8 deals, in part, with the perennial objection about the Malibu surfer as mentioned above. It will be argued, as noted, that the problem is likely to be that there are not enough perennial surfers, not too many. It may be necessary to radically reduce the work week to address the need for more jobs. In the longer term, capitalism itself will be shown to be inherently self-defeating. Capitalism, in the twenty-first century, will consume itself. The main driver will be technology, not social reform. One of the consequences of the robotic and computer revolution is that the price of owning the means of production will fall dramatically. Our task will be to work out a vision for a society and an economy that does not depend on motivating people to work through the threat of deprivation.

A short concluding chapter argues that BIG is not utopian. There are grounds for optimism for the possibility that we can adopt BIG in a peaceful and rational manner.

The next chapter discusses a very common objection to BIG: we simply cannot afford it even if we want to. In response, I make two proposals: one involves a new value-added tax of 14 percent on all goods and services; the other involves reforming the current tax code. The main point is that BIG can be paid for in a manner that will either help, or at least not financially harm, the vast majority of the citizenry. Even for the rich, the new tax rates would be well below what the highest rates have been in the past.

CHAPTER 2

Paying for Basic Income Guarantee

Introduction

A frequent objection to BIG, at least from those hearing about it for the first time, is that we simply could not afford to give everyone $10,000 a year. Indeed, in the minds of many, the United States has enormous yearly budget deficits. In actual fact, the budget deficit has been diminishing recently. In 2009, it stood at 1,400 billion. As of 2014, it was down to 483 billion. So the deficit has dropped by almost a trillion dollars in the past five years.[1] This is not to say that the deficit is no longer an issue; rather, given the amount of hyperbole out there, we should keep this in some perspective. Those predicting widespread rioting, looting, and cannibalism as a result of the federal deficit may have to wait awhile for their predictions to come true.

Still, even if the country is in better shape financially than some would have us believe, $10,000 per person is a lot of money. The question of where it will come from is a good one.

As a first stab at convincing you that this is not a matter of climbing Mount Impossible, consider that the Gross National Product of the US economy was over 17.5 trillion dollars, or about $53,000 dollars per person, in 2014.[2] The cost of BIG, as we shall see, is about 1.9 trillion dollars, which represents about 11 percent of the entire economy. Not a number to take lightly by any stretch of the imagination, but not an impossible number either. Most of the readers of this book could afford an 11 percent decrease in their pay without starving or going bankrupt. Yes, it would be painful but not impossible.

However, this is not what is being asked. Indeed, this chapter comes with glad tidings: the vast majority of US citizens would actually come out equal or ahead, financially speaking. Yes, taxes would go up for many, but so too would incomes. For the vast majority, a BIG payment would exceed any tax increase, and so most would be financially ahead.

This chapter sketches two proposals for financing BIG. The first proposal leaves existing tax policy just as it stands now and simply institutes a new sales tax to pay for BIG. As I will show, this new tax would actually be financially beneficial for most.

The second proposal is more involved. It tries to address some of the fundamental problems with the US financial house. Following Allan Sheahen's lead, we will show that about half the cost of BIG can be provided by simply eliminating most tax loopholes, and the rest can be made up by cutting waste and taxing the superrich more. This may all sound too good to be true, but it is not. We need simply to pull back the curtain on the plutocracy that benefit from the byzantine tax system and government spending at present. Our first task is to calculate the cost of BIG.

Calculating the Cost of BIG

As of 2013, there were 316,128,839 people in the United States. If each person received BIG, the total amount required would be over 3 trillion. However, almost 45 percent of this total would not be eligible for BIG under the present proposal. Once we exclude seniors, children, prisoners, and immigrants, the number of people eligible for BIG would be in the order of 188 million. In the following subsections, we will discuss these exclusions and get a ballpark estimate for the cost of BIG.

Seniors

The vast majority of the 44 million seniors in the United States already have access to something like BIG: social security.[3] In the future, it might make sense to combine social security with BIG, but for the present we will assume that seniors will either collect just their current social security or social security plus a supplement from the BIG fund to make a BIG equivalent, whichever is greater. An additional 49 billion dollars would be required to bring the lowest social security recipients up to the proposed $10,000 BIG.[4] So we will add 49 billion to our proposed budget. In other words, since social security is already in place, there is substantial savings in not having to provide a full BIG to all seniors. A full BIG for seniors works out to 440 billion dollars, but almost 90 percent of this is already taken care of by social security.

Prisoners

Estimates of the yearly cost of incarceration for the average prisoner in the United States are typically in the order of $25,000, plus or minus about $10,000 depending on the model used for calculation. In any case, the cost to society is enormous—much more than BIG on a per capita basis. The thought that society would have to pay to incarcerate criminals while those incarcerated receive $10,000 in their banks each year might be enough to drive even the most liberal liberals insane. One way to look at it is that rather than denying prisoners BIG, their BIG is being used to offset the cost of incarceration. Even the lowest estimates for the cost of incarceration are above what a prisoner would receive for BIG. For example, one estimate has the cost of incarceration in Louisiana at $13,000, compared to $45,000 in Rhode Island.[5] Indeed, there may be some symbolic value in the state charging prisoners $833 a month as their share of the cost of incarceration. Each month then, a debit and credit of $833 would appear on a prisoner's bank statement.

For the purposes of our analysis, prisoners are denied their BIG payment. This saves an enormous amount on the cost of BIG, since the United States boasts the largest prison population in the world. Indeed, the United States is a world leader and is internationally renowned for having 25 percent of the world's prison population, with only 5 percent of the world's total population: a frightening statistic. (If the United States changed its penal policies, it would only help the case for BIG, as such a change would free up substantial public monies.)[6]

Immigrants

BIG would be provided only to those who are citizens; so visitors, visa holders, and permanent residents would not be eligible for BIG. Permanent residents in need of financial assistance would have to apply for welfare. Again, I take this to be a conservative assumption. I take it that this assumption is necessary to allay the worry that the United States would be swamped by people looking to immigrate for "free money." For anyone to be eligible for BIG, he would have to go through the usual path to citizenship, which typically takes at least five years. There are now about 8 million immigrants in the 18–64-year range who would not be eligible for BIG.[7]

Children

Children are an important exclusion from the analysis.[8] There are approximately 74 million children under the age of 18 in the country. A full BIG payment to each would increase the cost by approximately 740 billion dollars. This is not, however, the reason they are excluded from the analysis. Rather, the question of giving BIG to children is so divisive that I believe it is better to deal with the issue more extensively on another occasion. I will limit myself here to a few general remarks that would need to be investigated more fully in subsequent work.

In general, there are two sorts of worries in providing BIG to children. On one hand, there is a potential for adults to use this as a means to enrich themselves. For example, imagine we are providing $10,000 per child. Two adults with eight children could then net $100,000 per annum. This could potentially encourage parents to have lots of children and use most of the money for themselves and neglect their children.

It is important to be clear about what is *not* being implied here. It is not suggested that BIG would encourage parents to have children for the wrong reason. A full BIG for children may encourage parents to have children for the wrong reason, but it is not evident that children would be significantly harmed by this action alone. Consider a woman who lies about taking birth control measures in order to get pregnant so as to keep her true love in her life. Let us grant, at least for the sake of argument, that this is not a particularly good reason to get pregnant. However, the child so born need not be significantly harmed by the mother's decision. If the parents are loving, clearly the child is not harmed. This is especially so when we consider the alternative is to not to have been born. The worry, if indeed it is a genuine worry, is not that BIG will promote adults to have children for the wrong reasons, but that it will promote *bad parenting*. To flesh out the example above, suppose a couple decides not to have children because they do not like children. However, when BIG arrives, they decide to have eight children to secure a large income for themselves from their neglected children's BIG. This would be a good case of BIG promoting bad parenting.

But this does not encompass all cases and would hardly decide the issue. We can just as easily imagine BIG promoting good parenting in many cases. BIG might make parenting easier and encourage better parenting by lowering financial pressures on a family. A detailed consideration of this aspect would require us to review the likely causes and consequences of a variety of current public policy choices on child rearing. Such a review is beyond the scope of this work.

The other extreme is to provide a very small BIG payment for children that it would remove the incentive to have children just for supplementing one's income through BIG. For example, in an earlier model of BIG, Allan Sheahen proposed a payment of $2,000 per child.[9] This is significantly lower than most estimates of what it costs to raise a child. This may not even cover food costs for ravenous teenagers. So, while this reduced amount may prevent prospective parents from having children just for financial gain, it would also put a heavy strain on single parents raising children using BIG as the sole source of income.

A third option would be to not provide any BIG for children, but instead use the traditional apparatus of the welfare state where needed. So, low-income parents can apply on behalf of their children for welfare. As with the current system, grant of welfare would be income-based. Persons making $200,000 annually would not be entitled to any welfare for their children, but a single parent living solely on BIG would be eligible.

A fourth option would be a combination of the second and third options. This would provide a small guaranteed income for each child. This amount would be small enough to keep parents from having children as part of some money-making scheme. Those earning low incomes could apply for welfare to boost this amount.

As noted, the issue of BIG for children is so important and potentially so divisive that it cannot be dealt with properly here. I take it that there is near universal agreement with at least this much: children deserve financial support either from their parents or the state, or from both. Even hard-line conservatives who think that the state should not support adults may agree for support to children. After all, the usual argument that adults should take personal responsibility does not apply to children. Since BIG is barely enough for a single person, it is clear that something must be done for children of single parents who live on BIG. In my discussions in this book, I will assume that the existing state apparatus will continue to look after children of impoverished adults. For example, there are existing programs to help poor families raise children, and these programs should continue to provide such support.

Let me say a little more about children in a strictly polemical mode: the reason I do not advocate BIG for children here is *not* that I am against children—I am a former child myself. Indeed, I think children should get *much more* state support than they do at present for at least two reasons. The moral reason is that children are economic innocents: they cannot work, nor can they be said to have consented to the present economic arrangements. Second, even from a cold, calculating, and investment standpoint, it makes sense for the state to invest more on children. Better-fed and educated children make for more productive adults. If you think I should say more about children, I do agree—but just not here. Children, it seems to me, are a very special case and must be dealt with separately. It may be that the best way to serve the interests of children is also through some sort of a BIG, or a graduated BIG. This remains an open question as far as I am concerned. In my mind, it is a huge mistake to think that adults and children can be simply lumped together for the purpose of policy making.

Total Cost of BIG

The exclusion of prisoners, immigrants, and children from our calculations for BIG serves the purpose of focusing on the clearest cases for BIG: law abiding, adult citizens. Nothing in the foregoing discussion makes a strong case for the exclusion of prisoners, immigrants, and children, so the present proposal should be taken as a *floor* rather than a *ceiling* in terms of who should receive BIG. That is, my claim is that, *at a very minimum,* we should include all (non-imprisoned) adult citizens. As noted, we will leave for another occasion the question of expansion of eligibility.

We are now in a position to calculate the total cost of BIG for a total number of 188 million eligible adults (see table 2.1).

A Value-Added Tax

One way to pay for BIG is to institute a new tax to cover it. The proposal would work for a new income tax as well, but I will explore the idea of what is known as a "value-added tax" (VAT) to pay for BIG.

Table 2.1 Total cost of BIG for 188 million eligible adults

Item	*Debit*	*Credit*
US Population		316
Children under 18	74	
Seniors (over 65)	44	
Green Card holders	8	
Prisoners	2	
Total population eligible for BIG		188

All numbers are in millions.

A VAT of 14 percent would generate sufficient money to pay for a $10,000 BIG policy. The proposed VAT would apply to absolutely all final goods and services. So, there would be a 14 percent tax on food, haircuts, books, medical services, nuclear missiles, and so on. However, consider those trying to live on $10,000 a year when the new VAT is introduced. Suddenly, their money would not go nearly as far. Imagine they spend $200 a month on food. After the 14 percent VAT is introduced, the same groceries would cost $228 a month. Their share of rent would increase from $300 to $342 a month. To have the same purchasing power as before the new tax, those living on $10,000 a year would need another $1,400 to pay for the VAT. So this is what I propose: increase BIG to $11,400. As should be clear, this will not make those living solely on BIG any better off, for although they have 14 percent more money, everything will cost 14 percent more because of the VAT. In other words, although recipients will receive more than $10,000, it will get reduced to $10,000 after paying VAT.

We calculated the cost of BIG at approximately 1.9 trillion without catering for the new tax. The higher BIG payment to offset the increased living costs due to VAT requires funding in the order of 2.2 trillion. As noted above, the gross domestic product of the United States (in 2014) was about 17.5 trillion dollars. A 14 percent tax on this value would yield 2.5 trillion in revenue to fund BIG (2.2 trillion) and also leave a small surplus (300 billion).[10]

One of the more controversial part of this proposal is applying VAT to all final goods and services in the economy. Generally, countries allow exemptions or reduced rates of VAT such that, for example, a 10 percent VAT might typically bring in about 4 percent revenue as a ratio of GDP.[11] This is referred to as a yield ratio, and for many nations the yield ratio is between 0.3 and 0.4, with Mexico marking out one end of the range at 0.28 yield ratio and New Zealand the other end, with 0.69. What this means is that if the 14 percent VAT is applied in a typical fashion, it would yield far less than is calculated. Even if New Zealand's high yield could be achieved, it would mean that a 14 percent VAT would fall short of the target 2.2 trillion by about half a trillion dollars. A VAT of about 18 percent would then be necessary to make up the shortfall.

The primary reason for applying a single rate of VAT across all sectors of the economy is to collect as much tax as possible. Inevitably, if countries decide to levy VAT at full rate to only some items and at a lesser rate on others, the tax base shrinks.

So, why do countries allow exemptions and reduce VAT on certain items? The answer is that there is a concern that VAT applied at an even rate across all goods makes it a very regressive tax. That is, unlike income tax, which tends to have a progressively higher rate for larger incomes, VAT is a flat tax, which tends to hurt the poor more. The objection, then, is that though it may make sense to have a VAT on luxury items, taxing basic necessities will hurt the poor. In other words, to help the poor, it is better to have a high VAT on yachts and a low or no VAT on food.

The counterpoint to VAT as a regressive tax is that the BIG dividend more than makes up for this for most people. In other words, most people will pay far less than $950 in VAT each month but will receive a $950 dividend check each month to make up for cost increase due to VAT. Focusing on the idea that the tax is regressive fails to take into account both sides of the equation: the tax debt and the BIG dividend. So it is possible to keep a large tax base and avoid the regressive tax criticism simply by topping up BIG to account for VAT for the poorest.

One area of this proposal that is particularly controversial is that VAT should apply to government purchases. Typically, governments excuse themselves from any kind of VAT. If the government buys a nuclear missile, for example, such a purchase would be classified as a final good or service, but it would be exempt from VAT. Government expenditures account for approximately 15 percent of the GDP, which works out to approximately $2.6 trillion.[12] This means that VAT on government expenditure works out to about $370 billion. These additional funds could be raised by either cutting many of the existing welfare programs, which will become redundant with BIG anyway (see below), or cutting the bloated military budget. If we did both, we would actually have a budget surplus (see below).

You Should (Probably) Love a New Tax

Of course, a new tax will raise additional worries. Who would want to pay an additional 14 percent tax on top of existing taxes? However, this question is misleading. The question, at least in self-interested terms, should be, "how much do I net?" rather than "how much do I pay?" I claimed earlier that most people would net more on this proposal; it is time to see this in more detail.

Suppose your salary is $50,000 per annum. A new tax of 14 percent would mean $6,100 in additional taxes. Yes, that is a lot out of a salary of $50,000. But remember too that you receive $11,400 in yearly BIG payments. So, under the proposal, your new total income is $61,400 (your $50,000 salary plus your $11,400 BIG payment).

If you consider your total income, your contribution to VAT would be $7,540, meaning that you would be actually earning $53.859, or nearly $4,000 more, on this proposal. The following graph illustrates the relationship between income, VAT, and BIG (figure 2.1).

It may help to walk through some of the information in the graph. The lighter dashed line represents the proposed BIG payment. It is constant for everyone: this is the by-now familiar point that BIG is exactly the same for everyone: from the homeless veteran, to Bill Gates. The darker dashed line represents VAT, which everyone pays. The smallest contribution to VAT is $1,400, which a person living solely on BIG would pay. Unlike the current income tax system, the proposed VAT is a "flat tax," meaning that the same percentage is paid no matter what a person's income. Someone making $10,000 a year pays 14 percent, as does a person making $100,000 a year, or even $100 million per year. The darker solid line represents a person's income in the present system. The lighter solid line represents the change in income (after paying VAT). In other words, the red line represents present income plus the BIG payment of $11,400 minus a 14 percent VAT.

The vast majority would do better under this proposal even though it includes a large new tax: anyone making between 0 and $81,000 a year would be monetarily better-off. Over 90 percent of the population has a net personal income that falls below the crossover point.[13]

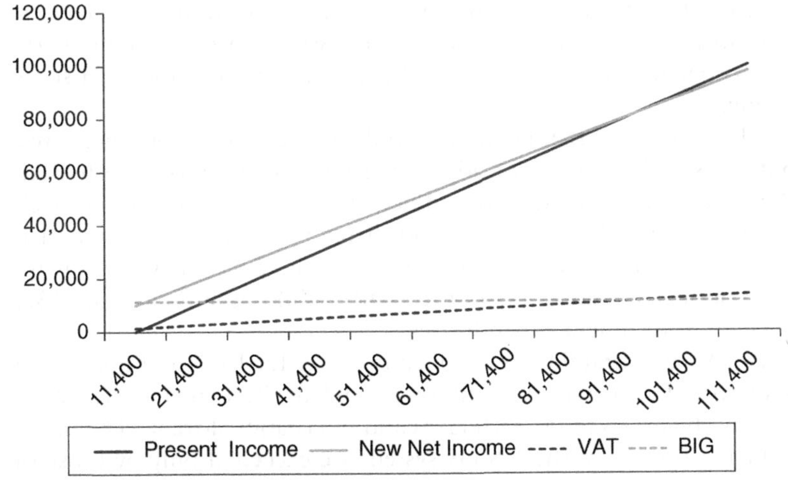

Figure 2.1 The relationship between income, VAT, and BIG

As noted, it is a very simple formula to illustrate that most would do better under BIG and that a 14 percent tax is hardly draconian. The top income tax rate for high income earners at present is 39.6 percent. So, the maximum that a higher income earner would pay under this proposal is 53.6 percent (39.6 percent income tax plus 14 percent VAT).

While this may seem like a lot, remember too that this is a maximum. The difference between the theoretical maximum and the actual tax paid (the effective tax rate) is enormous. The top 20 percent in terms of income earned pays only 20.1 percent income tax on average, and the top 1 percent pay a mere 20.6 percent.[14] In effect, then, the difference between the maximum and the actual income tax rate is about half. So, the present proposal for the top 1 percent would increase the effective tax rate to nearly 35 percent. Or to put it another way, if all the loopholes were closed and the richest taxpayers actually paid 39.6 percent, then they would be much worse off.

Again, this only applies to the super income earners. Consider how well someone making $100,000 a year would fare under the proposal. Looking at figure 2.1, we can see that they will fare less well, but the difference is almost imperceptible on the graph. The difference is actually $2,281 in additional taxes. This would only change the effective tax rate by 2.3 percent for those in this income bracket.

It may help to note that historically the highest rate for top earners was much higher. For most of the twentieth century the top rate was higher, reaching a peak during World War II at 91 percent. Historically, and though being average among the developed nations, US income tax is among the least progressive.[15] Again, I am not trying to defend my suggestion at this point, only putting it in some context.

I have assumed that nothing would change in terms of other taxes, but we should briefly consider this simple assumption. With BIG, there would be a greatly reduced need for income tax to support welfare programs. These monies could then be used to reduce the deficit or offset future increases in income tax. The amount, which, as we will consider below, works out to $400 billion, is sufficient to offset VAT on government expenditures.

A VAT is a straightforward means to pay for BIG, and instituting VAT would also assist with another problem the United States faces. This problem has to do with taxes and international trade. The sleep-inducing powers of the US tax system are exceeded only by those of international tax codes and agreements, so I will be brief. Basically, income tax is a tax focused on producers while VAT is a tax focused on consumers. European nations have VAT and income tax, but it will help to assume that all their taxes come through VAT. When exporting goods to Europe, US companies must pay tax as a producer and then, when the goods arrive in Europe, consumers are charged VAT. When Europeans export goods to the United States, they do not pay VAT, and Europeans do not pay American income tax. So, goods exported from the United States are taxed twice (once on American producers and once on European consumers) while European goods are not taxed at all. This is a greatly simplified description, but it is in essence the truth. If the United States introduced a VAT, it would put the country on a fairer playing field with the rest of the world in terms of international trade. There are only two reasonable options for rectifying the current situation. One would be to change international trade laws to alter how VAT and income tax applies to imports and exports. The chances of doing this are slim to none. As noted, the United States is the lone holdout that does not charge VAT, and so other countries are not going to be particularly motivated to change a system that now favors them. The only realistic choice is for the United States to change to the model that most other nations have adopted. As it stands, US companies are forced to compete with a heavy tax disadvantage; a situation that does not seem tenable in the long run.

A New Federal Budget

In this section, a second proposal for garnering funds for BIG will be explored. Instead of a broad new tax, this proposal involves probing the dank underworld known as the "US Budget." We will follow the late Allan Sheahen's lead here in reworking the budget to pay for BIG.[16] The basic strategy is to close tax loopholes, look for additional revenues, and eliminate unnecessary budget expenditures.

EliminateT axL oopholes

In terms of adjusting the tax code, the single largest way to contribute to the 1.9 trillion necessary for BIG will be to eliminate the many tax loopholes currently in play. This will result in an additional $820 billion in revenue to help finance BIG.[17] Before you sign on, you will probably want to know that there are 169 tax loopholes in the current system, and you are almost certainly a beneficiary of one or more of these. We will not go through them all here. I can tell you that, as a group, there is no rhyme or reason for their existence.

The IRS lists 169 "Tax Expenditures."[18] What the IRS calls "tax expenditures," the rest of us call "tax loopholes" or "tax deductions." What is the difference between a tax expenditure, a tax deduction, and a tax loophole? *Absolutely nothing*. Well, nothing other than whether one thinks the deduction is fair or not. In terms of the national budget, there is no difference. Every deduction or loophole claimed by one person must be paid by someone else.[19]

Here's an example of one of the ragbag of special interest tax deductions: "Temporary 50 percent expensing for equipment used in the refining of liquid fuels." If you do not know what that means, let me help you. I looked it up and it is a special tax advantage for petroleum refineries. Why the petroleum industry would need a special deduction when their corporate profit rates would make even Mafia dons blush is beyond me. In any event, this deduction put the country another $750 million in debt in 2011. With the present proposal, the petroleum industry will have to get along without this tax loophole.

More familiar is the homeowners' mortgage deduction. This is a deduction that at least makes a bit more sense. The deduction was instituted to help people buy their own homes. Although I can see the point of helping out low income owners to buy their first home, the majority of the deduction is not used for this purpose. Did you know that it is possible to write off interest payments on your 100-foot yacht if you live on it for more than two weeks in a year?[20] The homeowners' deduction added another $80 billion dollars of debt. And whatever good comes from it, we should still eliminate it on the principle of "no double-dipping." Most of the poor and the middle class will be better-off under our proposal. A low-income earner who gets the benefit of BIG will be much better-off than going without BIG and availing of the home interest income deduction.

Indeed, I am against all tax deductions, in principle, because they obfuscate how the tax burdens and benefits from the government are distributed. As mentioned, every deduction means someone else must pay. If there are certain positive results from the loopholes that we would like to remain, because we perceive them as socially beneficial, then they should be clearly marked as such. For example, at present, it is possible to claim a deduction for certain employer-provided day care benefits, and certain tax breaks are given to foster parents. If we want to retain these benefits, then the government should issue a check for the amount equivalent to the deduction rather than allow it as a tax deduction. In terms of the national deficit or surplus, this will not make a difference. But it will make a difference to public perception of the matter.

Here is how it would work. Suppose the tax deduction at present for employer-provided child care works out to $400 a year. On my proposal, you would pay the $400 extra in taxes, since there are no tax deductions, but receive a check from the federal government for $400 to offset the increased tax. Financially speaking, of course, this is a complete wash: paying $400 more in taxes and then getting $400 back from the government is revenue neutral. Similarly, suppose for some reason we think that the oil refining industry should get money from the federal government. Since there are no tax deductions, the industry would pay the full tax rate. So the refining industry will pay an additional $750 million. However, the federal government would issue a check to the individual industries to the tune of $750 million. Each check would have the heading "Corporate Welfare Provided by the Federal Government." Compared with the current system, the corporations sit exactly the same, financially speaking.

The whole point of this exercise is transparency. I am not suggesting that the refining industry should actually get this much in federal welfare. Rather, I believe it will make clearer to the taxpayers what is happening with federal revenues. I suspect people will be much more up in arms if they realize that the federal government provides $750 million in corporate welfare to the refining industry, rather than providing a tax deduction of $750 million. In some ways, this should strike us as surprising, since, as we have said, they are *exactly equivalent* in terms of revenue. All that changes is perception. And issuing a check with "Corporate Welfare" in the title better shines a light onto Washington where the interests of the plutocrats pull so many strings.

The total tax deduction works out to 1.025 trillion, and all of these will be eliminated on the present proposal. Our proposed budget takes only 80 percent of this revenue resulting in 820 billion in savings, meaning that there will be 205 billion that can be used to offset the increased tax rate for the most deserving. Exactly how this 205 billion will be spent will be open to negotiation. On the present proposal, however, this 205 billion will not be given as a deduction but as a payment from the federal government. Again, this will not make one iota of difference in terms of revenue, but it will increase transparency. Among the clearest cases for a tax deduction at present are deductions for the blind. Although the deductions would be eliminated on the present proposal, they would be replaced by a check from the federal government for the corresponding amount, which would come from the 205 billion pool of funds.

Elimination of Standard and Personal Deductions

Another $300 billion is saved in removing standard and personal deductions. One of the impetuses for these deductions was to help the working poor. The basic idea behind both deductions was to say, in effect, "we won't tax any income below subsistence level." For example, the standard deduction for single persons is $6,100, which means that the first $6,100 in income is not subject to income tax.[21] Of course, in practice, the standard deduction has not kept up with inflation, so now it is about half the official poverty line. In any event, with BIG, there would be no standard or personal deductions.

To work through an example, consider someone who can only find part-time, minimum-wage work and earns $6,200. Presently, he would pay income tax only on the last $100, so about $15 in income tax for a net income of $6,185. With BIG, every dollar in income will be taxed. So the same person would pay $930 in income tax. However, his total income would rise and it would net $15,270.[22]

Elimination of Different Tax Rates for Individuals and Couples

Presently, there are different ways of reducing income tax for families. The categories are "Married Filing Jointly," "Married Filing Separately," and "Head of Household." An additional $60 billion is raised by eliminating these tax loopholes.[23] The reason this is more than fair is that BIG is paid on an individual basis. Certainly, it would not be fair if BIG were paid per family, that is, if each family received $10,000. So, a married couple would receive $20,000, that is $10,000 each, under BIG. Thus, overall, middle and low income couples will still benefit financially even when these tax loopholes are eliminated.

Uniform Payroll and Income Tax

An additional $220 billion will be raised if the 12.4 percent payroll tax is applied uniformly. Presently, it magically disappears for incomes above $106,800. If you wonder why it suddenly disappears at this level, ask your favorite plutocrat.

Hiking the capital gains tax from 15 percent to 35 percent will raise another $88 billion per year. The fact that capital gains are taxed at such a low rate is one of the more bizarre tax decisions, at least from the perspective of fairness. After all, as Warren Buffett points out, it seems bizarre that those performing mental and physical labor

Table 2.2 Additional revenues by bringing changes to US tax policy

Item	*Debit (in billions)*
Eliminate 80% of 169 tax loopholes	820
Standard and personal deductions	300
Elimination of joint filing	60
Uniform Payroll Tax	220
Capital Gains Tax raised to labor rate	88
20% Surcharge on income over $1,000,000	129
Total revenue increase	1,617

should be taxed at a higher rate than capital gains; capital does not toil or sweat.

Surcharge on Incomes Over One Million

A 20 percent surcharge on income over one million would generate $129 billion in revenue. Do not worry, this tax does not apply to you. It does not even apply to most of the so-called 1 percent. In fact, 90 percent of the elite 1 percent would not be affected by this surcharge, since only 0.1 percent of individuals or households earn over one million in income.

Total Additional Revenues

Table 2.2 summarizes the additional revenues by bringing some much needed changes to the convoluted world of US tax policy. The additional revenues are almost enough to pay for BIG without any further changes.

Budget Savings

In this section, we will propose two major cuts to the US budget: welfare and military payments.

Cut Welfare

There are over 100 federal programs to help the poor, including food stamps, unemployment insurance, and make-work pay tax credit. These add up to over $460 billion.[24] Most of these will not be necessary with BIG. We propose to cut $400 billion from the budget. This will still leave $60 billion for those not covered by BIG and those who may need additional assistance.

End the Empire: Cut the Military

President Eisenhower was incredibly prescient in proclaiming the dangers of the rise of the military industrial complex. Consider that much of the current budget shortfall is due to the huge expansion of the US military budget. It went from $265 billion to $700 billion in just over a decade. It has recently inched down to $640 billion. BIG will be paid for, in part, by cutting back the budget by $300 billion to the year 2000's level of funding.[25]

It may be objected that such a drastic cut would leave the country vulnerable to attack. Let me make two responses to this.

First, with funding at $340 billion, the budget is still huge in absolute and relative terms. Consider that a budget of $340 billion is equivalent to the combined budget of the next three largest military spenders (China, $188 billion; Russia, 88 billion; and Saudi Arabia, 67 billion).[26]

Second, a reduction in the military budget would make the country safer. This may sound paradoxical to many Americans, but to the rest of the world, the point is probably too obvious to make.

Ostensibly, the large build up in the American military since 2000 is to fight the war on terrorism and the wars in the Middle East. When such huge amounts are being spent, you would think the general public would be much more interested in whether the funds are achieving their desired ends. Consider this: I have asked many of my students, "why do terrorists hate the United States?" Quoted here is a typical answer that, if graded, would fail:

> They hate us, not because of anything bad we've done. This has nothing to do with any aggression on the part of the United States of America. It has nothing to do with anything America is taking from anyone. It has nothing to do with Israel and Palestine. They hate us for the freedoms that we have and the freedoms that we want to share with the world, because the freedoms that we have and the freedoms that we want to share with the world are in conflict with their perverted interpretation of their religion, their maniacal, violent, and perverted interpretation of their religion in which they train their young people to be suicide bombers, and they train them to hate you and despise you, and they train them to hate your religion, and to not allow you to have religion of your own, or anyone else. They hate us for the reasons that are the best about us, because we have freedom of religion, because we have freedom for women, because women are allowed to participate in society, because we have elections, because we have a free economy. Well, we're not giving that up, and you're not going to come and take it from us.[27]

This is Rudi Giuliani's take on the reason for terrorism. The former mayor of New York uttered this nonsense while running for the Republican nomination for president. The overwhelming consensus of academics who have studied the causes of terrorist attacks against the United States is that primarily terrorists get incited to react against what they see as an Imperialistic American Empire.[28] The thought that this is merely the opinion of left-leaning academics is belied by the comments by another Republican candidate:

> The reality is that our military presence on foreign soil is as offensive to the people that live there as armed Chinese troops would be if they were stationed in Texas. We would not stand for it here, but we have had a globe straddling empire and a very intrusive foreign policy for decades that incites a lot of hatred and resentment towards us.
>
> According to our own CIA, our meddling in the Middle East was the prime motivation for the horrific attacks on 9/11. But instead of re-evaluating our foreign policy, we have simply escalated it. We had a right to go after those responsible for 9/11, to be sure, but why do so many Americans feel as if we have a right to a military presence in some 160 countries when we wouldn't stand for even one foreign base on our soil, for any reason? These are not embassies, mind you, these are military installations. The new administration is not materially changing anything about this. Shuffling troops around and playing with semantics does not accomplish the goals of the American people, who simply want our men and women to come home. 50,000 troops left behind in Iraq is not conducive to peace any more than 50,000 Russian soldiers would be in the United States.[29]

These are the words of Republican Presidential candidate Ron Paul. Ron Paul's point here seems pretty obvious to the rest of the world. But it is pretty hard for many Americans to parse: many people object to American troops in their country.

To be clear, I am not suggesting that the way to make America safer is to reduce the military budget—at least not directly. I am saying that the way to make America safer is to reduce its military footprint: reduce the huge global presence of the American military and stop throwing the US military might around. A smaller global presence will require a smaller budget. In terms of policy, it might help to simply cut the budget. This would force the military to reduce its presence overseas. Our military's massive overseas presence is what makes us so many enemies and what drives up costs for taxpayers.

America needs help, it is a junkie hooked on war. The military-industrial complex that Eisenhower warned us about is the primary beneficiary of a huge global empire. After the Vietnam experience, many Americans are hypersensitive about supporting men and women in uniform. But it is hardly a subtle distinction to say: "I support the men and women in arms but not the agenda of the military-industrial complex." Indeed, since it is the agenda of the military-industrial complex that gets so many soldiers needlessly killed overseas, to effectively support the former requires being against the latter. Any action that requires the sacrifice of American men and women in combat should have proper representation from all strata of society: 1 percent of the US military should be made up of folks from the 1 percent wealthiest portion of the population. Until then, there is a serious worry that the poor are being used as cannon fodder for the military-industrial complex. As with any junkie, the first step to recovery is admitting there is a problem.

TotalB udgetS avings

By cutting welfare and the military budget, we yield $700 billion in savings (table 2.3).

New Budget

The surplus is $377 billion, which is about what is needed to pay for this year's projected deficit under the old budget. So the proposal would yield close to a balanced budget (table 2.4), given the projection of a $483 billion dollar deficit in 2013.[30]

Table 2.3 Total budget savings

Item	*Debit in billions*
Welfare cuts	400
Military cuts	300
Total budget saving	700

Table 2.4 Newb udget

Item	*Debit in Billions*	*Credit in billions*
Cost of BIG		1,930
Additional revenues	1,617	
Budget cuts	700	
Subtotals	2,317	1,930
Total cost (surplus)	387	

Universal Health Care

I have kept the biggest savings for last. Let us follow suit and use the rhetoric of the American right and call government funded health care "socialized medicine." On a per capita basis, the United States spends more than almost all other economically advanced nations (other than Norway and the Netherlands) on socialized medicine.[31] So the country is a world leader in socialized medicine: the US government spends more on health care on a per capita basis than countries with universal health care such as Canada, the United Kingdom, France, Germany, and Italy. This ought to strike us as very puzzling: how can the United States spend more per capita on socialized medicine when it does not have universal health care? That is, when many citizens are covered by private insurance or no insurance at all?

It is a fact that the United States spends so much more on health care on a per capita basis than any other nation. The average of the developed nations is 9.5 percent of gross domestic product.[32] The United States spends nearly twice that, at 17.6 percent.[33] The proposal here is to model the US health care system after Canada, which spends 11.4 percent and provides universal health coverage. The savings to the economy works out to $893 billion dollars. (US gross domestic product is $14,419 billion and a Canadian-style system would save 6.2 percent, which works out to $893 billion.)

Note that it would be widely utopian to think that these saving would be realized the first year. Switching from the present system to a Canadian system would have all sorts of legacy costs. For example, thousands of administrative positions would be eliminated as all medical billing is streamlined into a single payer system. It might take five or ten years before the full savings of universal health care are realized.

But what about the quality difference in health care between the two systems? Some argue for the superiority of universal health care systems in terms of measures such as infant mortality and life expectancy. The United States lags in these statistics compared with countries with universal health care. The country also lags in terms of "amendable mortality"—deaths that could have been prevented with medical treatment.[34] Defenders of the US system will point to possible social factors that might explain these differences: violence, poverty, and obesity, for example.

The most common criticism of the Canadian system by apologists for private medicine is the "wait times." Here the old adage that there are three types of lies seems appropriate: "Lies, damned lies, and statistics." To examine the problem, let us work out an example. Suppose we randomly sample 100 people needing hip replacements in both Canada and the United States. In Canada, let us suppose the average wait time is six months. In the United States, 90 patients wait three months, and 10 never get the hip replacement because they lack insurance. The average wait time for those who get surgery is lower, but how do we factor-in the last ten. We could just ignore them and then say that the average wait time in the United States is half that of Canada. Or we could add up the number of months the patients live and then consider that as part of their "wait time." Say the ten patients live another ten years each, on average. Then the average wait time in the United States jumps to 14.7 months. Or we could say that since these ten never receive hip replacements, their wait time is infinite, and so the average wait time in the United States is infinitely long. If we include those that are not treated at all, then clearly the wait times are much longer in the United States. My point is not that any of these measures are in fact correct, but that there are clearly important assumptions about who is counted and how.

Although many of the commonly used measures for evaluating national health policy favor the Canadian system, the differences in the system between countries are quite small and could easily be explained by causes other than differences in health policy. The two most significant differences clearly favor the Canadian system. First, the Canadian system covers everyone. Second, it does so much more economically, the US system costs nearly a trillion dollars more. There is absolutely no indication that the additional trillion dollars buy the average American significant benefit. Admittedly, it does serve the interests of some, including the wealthiest. So, no single health system will appeal to the interests of everyone. Hard choices must be made: should we have a system that benefits the majority or the wealthiest? The present system, even in its revised "Obamacare" form, favors the latter.

It may be objected that if we opt for a Canadian-style health care system, taxes would have to go up dramatically to finance the increased cost. The answer to this is the mind-blowing statistics I detailed above. *The US government already spends more than Canada per person.* I say this is mind-blowing for obvious reasons: private health care currently costs many middle- and lower-class Americans a huge chunk of their salaries. Americans could have universal coverage modeled on the Canadian system and taxes would actually *decrease.*

Conclusion

It should be emphasized that the aim of this chapter is to show that it is, in principle, possible to generate revenue to pay for BIG. When the time comes to get down to the nitty-gritty of actually changing tax policy and government expenditure, it would be wise to weigh more options than just the two considered here. After all, there are many permutations and we should sort through these carefully. For example, a higher VAT of 25 percent, like that of some European nations, on a smaller section of the economy might work better than 14 percent across the board. Another possibility would be to combine the two proposals, for example, a smaller VAT, say 8 percent, with the plugging of fewer tax loopholes.

So, the intention of this chapter is not to offer something like a blueprint for reform. A serious effort at reform would no doubt involve dozens of economists and policy analysts considering dozens of proposals regarding the best way to finance BIG. These analyses would probably fill a book several times the size of this one. What is offered here is something far less ambitious: a "back of the envelope calculation." The point to be emphasized is simply that there is absolutely no economic reason that a BIG could not be financed. Recall that it represents only 11 percent of the entire economy. Politically, it might be difficult to get the changes passed, but that is a different sort of objection (which we will consider briefly in chapter 8).

CHAPTER 3

Fulltime Capitalism: Basic Income Guarantee as a Dividend from State Capital

Introduction

So far, I have described BIG and argued that we could easily pay for BIG if we chose to. Most of the remainder of the book is an attempt to persuade you that you should support BIG.

For many readers, one argument for BIG is implicit in the previous chapter: BIG is in your self-interest. As noted, BIG would improve the financial situation of most people. Of course, if your income is in the top 10 percent or so, then in purely self-interested terms, you ought to be against BIG. After all, the current sociopolitical situation has been carefully crafted to your advantage. Why change a good thing?

It is not clear how much stock we should put in an argument that appeals to self-interest. Consider that there is a long-standing debate about how voters should vote: should you vote what is best for you or best for your country, as Rousseau thought?[1] As is often noted, voting patterns are perplexing if voters are assumed to vote for their economic self-interest. Many poor areas of the United States vote Republican and many rich areas vote Democrat. This seems to belie the fact that the poor tend to do better economically under Democrat fiscal policy rather than Republican. There are many possible explanations for this voting pattern, one among them is that people vote for what they think is best for the country rather than what is best for themselves.

Fortunately, we do not need to decide the question of whether people should vote in a self-interested fashion or in what they perceive is the best for the country. As we shall see, both oars pull in the same direction.

This chapter argues that a good case for BIG can be made on the basis of capitalism. This may seem surprising, since in chapter 1 we noticed that BIG is often thought to be a compromise between capitalism and socialism. The thumbnail version of the argument that capitalism can be harnessed in support of BIG is as follows. We are all rich stockholders in state capital. A dividend from this wealth is sufficient to provide us with BIG. The money that should go to BIG in the form of dividend actually subsidizes the rich in this country. In other words, from a purely capitalistic perspective, we should see that the poor are subsidizing the rich.

Put another way, the argument of this chapter is that citizens of the United States are asset-rich. This will no doubt come as a surprise to poor minimum-wage earners, students struggling to decide whether to buy food or books, the unemployed desperate for any job, homeless people eating out of garbage cans, and so on, but it is true. Consider someone who has a valuable asset, such as land or stocks, but is cash-poor because she is not making an income off her assets. What this person needs is a way to leverage her assets to make an income. Similarly, American citizens own high-value assets in the form of state capital. As good capitalists, they should seek to make an income from these assets. This chapter is a call to make more income from state capital.

Capitalism

The first thing we need to think about is capitalism itself. Typically, capitalism is thought of as the combination of two components: private ownership of capital and a free market where goods are bought and sold. The following definition is fairly standard in wedding these two ideas in defining capitalism:

> An economic system characterized by private or corporate ownership of capital goods, by investments that are determined by private decision, and by prices, production, and the distribution of goods that are determined mainly by competition in a free market.[2]

The two ideas are separable. Imagine a small agricultural country where extended families own farms. There is a strong tradition of economic independence. Each family does as much as possible to be self-sufficient. Any trading with other farms is done only in emergency situations and results in a loss of status for the farm initiating the trade. In such a country, there is private ownership of capital—the farms and farm equipment—but little free market. Alternatively, we can imagine another country where the king owns all the businesses, land, and equipment in his kingdom. He has businesses compete against each other in a ruthless free market. The king rewards handsomely the managers and workers who do well in the free market, that is, those who turn above-average profits. The managers and the workers of the least successful business are publicly flogged. The king sits back and delights in the cutthroat competition. Wealth earned by employees can be used on personal luxury goods, but anyone attempting to start a business to compete with the king's assets faces the prospect of summary execution. In both countries, capitalism as we know it is absent: capitalism requires both the private ownership of capital goods and free markets.

Marxism

Our next step is a very cursory understanding of Marxism. But first a caveat: one way to try to keep American children in line is to tell them that if they do not behave, Karl Marx will get them. Good bourgeoisie parents will check under their children's beds for Marx to reassure anxious American children at bedtime. Yes, this is a bit hyperbolic, but it is no exaggeration to say that Marx is the bête noir of American politics. True, many bad things were done in the name of Marxism; for example, the Soviet system of industrialized death and the Cultural Revolution in China took tens of millions of lives each. Of course, many bad things have been done in the name of Christianity too. If we do not blame Christ for the genocides committed in his name, then surely we must ask why the same reasoning does not apply to Marx. I am not saying that you should believe in Marxism. I am saying that many readers in the United States and other parts of the world need to throw off a lot of indoctrination to consider Marxism logically and dispassionately.

Let us think first a little about coercion: suppose you are held up at gunpoint by a thug who says, "your money or your life." Do you have a choice in this situation? Yes, in one sense: you really can choose between your money and your life. (Let us assume that you have good reason to believe the thug will not steal from your dead body. He is from a rare breed of honest murderers.) But you do not have a choice about whether you are subject to the coercion of the thug. In other words, you have a free choice *within* this coercive relationship imposed by the thug, but you do not have a free choice *about* adopting the coercive relationship.

Marx said similar things about capitalism. In a capitalistic economy, workers are free to take a job offer from another employer. In important respects, then, capitalism offers economically lower classes more freedom than earlier economic systems; for example, neither slaves nor serfs were free to take up job offers from different masters or lords. So, although capitalism offers greater freedom for the economically disadvantaged classes than slavery or feudal societies, it is, according to Marx, still coercive, since workers are not free to choose to work in a noncapitalistic society. In other words, workers are free *within* capitalism but not *about* adopting capitalism.

Why would workers want to choose something other than capitalism if it is a freer economic system than slavery or feudal society? The short answer is, according to Marx, even freer societies are possible: Marx thought that some forms of socialism and communism offered even more freedom.

According to Marx, private ownership of the means of production permits capitalists to operate like vampires sucking time out of workers. The point may be illustrated with a simple example of two identical factories in two different economies. The pay at factory A in a capitalist economy is $10 per hour. A hundred workers receive this wage from the factory owner. The factory nets $15 per hour, so at the end of the eight-hour workday, the capitalist earns $4,000. Factory B is in a socialist economy and is owned by all those who work there. Their pay is $15 per hour and so there is no surplus profit to distribute.

Marx claims that workers are freer in factory B than in factory A for the most obvious of reasons: their additional earnings translate into more freedom. For one thing, if the workers in factory B were content with earning $80 per day like their counterparts in factory A, then they would need only work 5.33 hours, meaning they would have 2.66 hours more free time every day. Or they could choose to work 8 hours like those in factory A and use the additional income to make monetary choices that would be beyond the reach of those in factory A on a lower income. So, what the capitalist will describe as making a profit, Marx will describe as the siphoning off of time and freedom from workers.

The capitalist will dispute the Marxian analysis here by pointing out that no one forces workers to work in factory A. They are there of their own free will and so there is no sense in which capitalism is coercive or analogous to slavery or feudal society. But, as intimated above, workers are not offered the choice of working for capitalists or adopting a socialized model of ownership of the means of production.

Rather, the choice the average worker faces is to work for capitalists or starve. In other words, the Marxian analysis is that capitalists offer this choice to workers, your life or a portion of your time and freedom, the same choice offered by the thug.

This relatively simple point seems to go unappreciated by many authors. The following quote from the eminent professor of law, Richard Epstein, is emblematic of how to misunderstand Marx on this elementary point:

> To the determined Marxist or his modern sympathizers, however, exploitation often carries the more cynical connotation that one side of the transaction is left worse off than he would have been if he had never entered it at all.
>
> But this view of exploitation offers no explanation as to why someone down on his luck would choose to make a contract that left him poorer than before. Many contracts are performed on a repetitive basis: the ordinary worker can quit at any time and yet frequently will return to work day after day. He obviously does so because he thinks that this opportunity is better than any of his alternatives, and it would be an odd form of assistance to ban him from that line of work altogether.[3]

Marx would agree with Epstein here that *within capitalism* the worker is better-off to repeatedly contract with an employer, given that the alternative is to be penniless and perhaps living on the street eating from garbage cans. So, yes, the worker is free to quit and free to eke out an existence on the street. Similarly, if I am held up at gunpoint repeatedly, it is always a good choice to give up my money and save my life. My complaint in this instance is not that I am left worse off by having my life spared as I am relieved of my money. Rather, the complaint is that I am forced to make such a choice. Similarly, Marx says that workers are forced to make a choice within capitalism: either have capitalists skim some of your labor or starve. Notice that Epstein is making comparisons about how choices are advantageous *within* a capitalistic system, which is to say that he has entirely missed Marx's point. Marx claims there are economic arrangements that make for greater freedom for workers than what capitalism offers. To point out, as Epstein does, that there is a certain amount of freedom within a particular institutional arrangement, that is, within capitalism, is the most facile understanding.[4]

It is worth emphasizing at this point that I am not saying that Marx is correct or incorrect. Rather, the point is this, we need to at least understand Marx better than Epstein does before making any suchj udgments.

Inverted Marxism: A Purer Form of Capitalism

Marx makes two important claims: socializing ownership of the means of production would lead to increased freedom for workers, and capitalism is inherently unjustly coercive. Both claims have been challenged. A common response to the first claim is that workers would not be (or are not) freer under a socialized ownership of the means of production. Proponents of this view might point to the inefficiencies of socialist economies. Workers often have to work as many hours as their counterparts in capitalistic economies for less money, and so Marx was wrong to think that socialized ownership of the means of production would result in more freedom.

The second claim, that capitalism is inherently unjustly coercive, has been challenged in two ways: it has been argued that private ownership of the means of production is coercive, but not unjustly so, and that the private ownership of the means of production is not coercive.

For present proposes, we will follow suit with the critics of Marxism and reject both claims by Marx. That is, we will accept, at least for the purposes of argument, that socialism will not lead to more freedom for workers, and that private ownership of the means of production is not unjustly coercive. What I want to suggest is that neither is sufficient to vindicate the contemporary version of capitalism. To show this, I propose to turn Marx's argument on its head. We should consider what it would mean to privatize *all the means of production*.

To see why it is plausible to think that not all the means of production have been privatized, we should attend to the fact that capitalism comes in degrees: countries can be more or less capitalistic. One dimension of particular interest is the degree to which one can attempt to use one's capital, that is, one's assets, to generate a profit. So, for example, two countries are otherwise identical but country C allows its citizens to make a profit on owning real estate, whereas country D does not. In this case, we should say that country C is more capitalistic than country D. We will denote this by saying capitalism is a gradable concept; economies can be more or less capitalistic. Not every concept is gradable in this way. A classic example is pregnancy, pregnancy does not come in degrees.

The gradable aspect of capitalism is further confirmed when we note that profits might be limited and prohibitions might exist on making profit on some assets in certain ways. As an example of the former, suppose country E is just like country D, but then attempts to become more capitalistic by lifting the prohibition on making money on real estate. People renting homes start complaining to their government representatives as their rents go up, so the government puts in rent controls, which limits the amount rents can rise in any given year. Rent controls limit the amount of profit that real estate capitalists can generate, still, country E is more capitalistic than country D, since it allows at least some profit to be made on real estate. However, E is not as capitalistic as country C.

Countries often allow particular forms of capital to be used for some profit-making purposes, but not others. The ownership of a gun might be part of the business assets of a guide who takes rich people on hunting trips. The gun, however, cannot be used as part of a hit man service. So, there are always limits on how assets might be used to generate profits. A more recent example is the change in policy about using automobiles to generate profit. It has long been the case that one can use one's car as part of some businesses in an attempt to make a profit, for example, real estate agents use cars to drive clients around to look at homes put for sale. Some jurisdictions lately have lifted bans on using private automobiles as taxis. Uber matches people wanting rides with drivers willing to offer rides for a price. Jurisdictions that allow Uber to compete with taxi services are purer forms of capitalism (other things being equal) because they allow one to use one's capital (in this case, one's car) in more ways in attempts to generate a profit.

We can generalize this by saying that in any comparison between two countries, or two different periods in one country, X is a purer form of capitalism than Y if X has fewer restrictions on attempts to make maximal profit on capital than Y (other things being equal). Note that this is not equivalent to saying that a purer form of capitalism leads to a better society or a better economy. Some have argued, for instance, that allowing people to use their cars as taxis leads to bad consequences, such as a negative effect on the livelihood of taxi companies, or an increased risk for paying passengers. If these arguments hold sway, then they are arguments against moving to a purer form of capitalism.

State Capital and Part-Time Capitalism

Let us think of "part-time capitalism" as the economic system that combines capitalism with noncapitalistic elements over a large class of assets. For example, a country that allows capitalism to flourish for smaller firms, say those with fewer than 20 employees, but not larger firms, which are owned by the state and managed for the good of the people. This is a form of part-time capitalism. The argument I hope to make is that the United States and other industrially developed nations are forms of part-time capitalism because there is a large class of assets, namely, state assets, that are not managed in a capitalistic fashion. The proposal is to move to capitalism full-time where citizens seek to make a profit on all of their capital.

So, two claims are being made: (1) the United States has an enormous stock of state assets, and (2) these assets are not managed in a capitalistic fashion. An easy way to illustrate these claims is to think of national highways. National highways are owned by the United States, but, for the most part, they are not managed in a capitalistic fashion. Consider that, for the most part, the use of highways is given away freely. The point is perhaps easier to see in contrast: some highways are privately owned. On a privately owned highway, one typically pays a toll and the owner of the highway hopes to make a return on the ownership of the highway.

The argument we will make is that the United States owns far more in the way of state assets than is generally realized, and that these assets could be leveraged to generate BIG. We will make the argument by analogy by first drawing some lessons about a publicly traded company, eBay.

The eBay Analogy

Paradigm cases of the "means of production" in Marx's corpus are land and giant factories billowing smoke. These paradigms are, of course, a function of Marx's time: they were some of the most conspicuous examples of capitalism in the nineteenth century. For Marx, however, capitalistic exploitation is not limited to the manufacturing sector of the economy. Rather, the important point is that the conventions of economic distribution embodied in capitalism are such that they permit private ownership of property to exert the aforementioned vampire effect: sucking labor and freedom out of workers. Thus, the point remains even if we turn from the manufacturing industry to the service industry. For example, if we change the previous example to one where the capitalist owns a string of hair salons and pays the workers $10 per hour while netting $15 an hour, nothing changes in terms of Marx's analysis. The private ownership of the salons allows the coiffeur-capitalist to exert the vampire effect.

Now consider eBay. eBay offers its customers an efficient online market for buyers. It generates income in various ways, including transaction fees. These fees start at about 3 percent for items listed. eBay is one of the darlings of the business Internet boom. It is valued in the tens of billions but, like many twenty-first-century companies, most of its value does not lie in the ownership of Marx's archetypical means of production: billowing factories, large tracts of land, railways, shipping lines, and the like.

Accountants make a distinction between tangible and intangible assets. Examples of tangible assets besides Marx's archetypical means of production are things like office chairs, computers, vehicles, equipment, and so on. Intangible assets are sometimes referred to as "nonphysical" assets. In saying that they are "nonphysical," typically accountants do not mean to take a stand on the deep metaphysical issue about the ultimate substance or substances of the universe—they are dull accountants, after all. Very roughly, by "tangible assets," accountants mean something that you could put a barcode on for inventory purposes, which is not generally possible with intangible assets. It is easy to imagine putting a barcode on land or vehicles owned by eBay, it is much more difficult to think about how to put a barcode on its proprietary software or the goodwill that eBay enjoys. Most of eBay's value lies in its nonphysical assets, in particular, its goodwill.

eBay is, by far, the largest online auction site. It seems that no one knows for sure how many online markets there are, in part because the basic software to coordinate buying and selling is fairly easy to set up. Clearly, it is not the computer program or eBay's computer servers that explain its value and stock price. Rather, the most important asset of eBay is its reputation and brand identity. Buyers find a large market of sellers, and sellers find a large buying pubic. Both feel that the eBay forum provides a certain amount of security to their economic transactions. Buyers can look at feedback on prospective sellers they are interested in doing business with and know that if the transaction does not go according to the terms specified, they can punish the seller with a negative review. eBay also offers dispute resolution mechanisms for buyers and sellers. It is not too misleading to summarize this by saying that eBay sells a huge "market of trust" to its customers. So, eBay's primary means of production is intangible: it is far more abstract than the factory billowing smoke. The temptation to call the goodwill eBay enjoys a nonphysical asset is easy to see. It is hard to point to something physical in the world and say, "this is what eBay owns that makes it so valuable." Nevertheless, the reality of what eBay owns is undeniable: we see evidence of it in its ability to generate enormous amounts of income.

Naturally, shareholders in eBay do not provide the online market out of the kindness of their hearts; rather, the point is to make a profit. Of course, in a capitalistic society, this is to be expected. They own the means of production and are entitled to make a profit if they are rewarded by the market.

Now consider this thought experiment: eBay stock is bought from its present owners and new stock is distributed to all adult citizens in the United States, one share each. To make the math easier, we will assume that this works out to exactly 200 million adult citizens in the country, and so there are 200 million eBay shares. (A consequence of our assumption is that there is no non-US ownership of eBay.)

As a stockholder, you might ask yourself: how should eBay be managed? The question is potentially ambiguous. It might mean how should eBay be run to best serve your personal interest, or the question might be about how to best run eBay to maximize profits for its shareholders. It may be thought that these two questions always yield the same answer: anyone looking to maximize his or her profits ought to try to maximize the profits from any particular asset. While this is often the case, I will show that this is not always the case.

In thinking about how to maximize your earnings, it will help to know a little bit more about how eBay works. eBay has a byzantine system for making money. It charges sellers to list items up for auction and it charges a percentage of the selling price of an item using various scales and formulae. For our purposes, we can simplify and assume that eBay makes all its revenue by taking 3 percent of the sale price of an item. We will call this the "transaction fee." We will suppose that eBay's net income is 2 billion a year, which is not too far from the actual figure.[5] We will also suppose that eBay distributes this income to its shareholders. Based on our assumptions then, each shareholder should receive a $10 dividend for 2012.

Let us suppose that shareholders find themselves having to vote for one of two candidates for CEO of eBay. Bush runs on a platform of lowering transaction fees to 2 percent, while Xram proposes to raise transaction fees to 4 percent. Some worry that Bush's proposal will lead to eBay being a lot less profitable. Proponents of the Bush transaction-fee cut argue that the profitability of eBay may go down, but this will stimulate market activity, particularly among the eBay high-income earners, and so everyone will be better off. Proponents of Xram's proposal claim that eBay is underperforming because the market will easily bear a higher transaction fee.

To keep things simple, let us suppose that leading economists agree on the following: at a 2 percent transaction fee, eBay will no longer have any net income but total sales will increase by 10 percent. At a 4 percent transaction fee, net income will rise to $3 billion and pay a dividend of $15 per share. (We will assume that gross sales will go down by 10 percent, but profitability will be up.) How should you vote if you want to make as much money as possible? The question may seem a no-brainer: you should vote for Xram's proposal. After all, this will return the highest dividend to you as a shareholder. If your only source of income is the stock dividend, then this is the correct answer. But if you use eBay as well, then the answer is not so simple. For you will have to factor the rate of the transaction fee into your overall financial picture. In other words, the answer will depend on how big a player you are in the eBay economy. Table 3.1 illustrates the relative overall expected income, factoring in the stock dividend and the transaction fee. The first number in each cell represents the stock dividend of a given transaction fee and the second number is the money that must be paid in terms of a transaction fee for a given yearly sales number. Let us think of eBay "low-income earners" as those who sell $0 to $100 per year. Clearly, they benefit by adopting Xram's proposal, for at $100 yearly sales, they pay $4 in transaction fees and receive $15 in stock dividends, netting $11. The eBay "middle-income earners," those who sell between $200 and $300 per year, also benefit from Xram's proposal. At $300 yearly sales, the transaction fees owed ($12) are almost offset by the dividends paid ($15), yielding a surplus of $3. The eBay "high-income earners," who sell more than $800 a year, should vote for the Bush proposal if they want to maximize their personal income. For on $1,000 yearly sales, they would owe $20, while on Xram's proposal, they would owe $25. The most financially advantageous strategy for each level of income is in bold.

To summarize, those who make the most use of eBay, the eBay high-income earners, should endorse Bush's proposal if they want to maximize their earnings. Those who make little or no use of eBay for buying and selling, the eBay "low-income earners," should endorse Xram's proposal to maximize their economic gains.

Table 3.1 Relative expected income, factoring stock dividend transaction fee

Transaction fee	*Yearly sales on eBay*					
	0	100	200	300	800	1,000
Bush 2%	0+0=0	0–1=–1	0–2=–2	0–3=–3	**0–8=–8**	**0–10=–10**
Status quo 3%	10–0=10	10–3=7	10–6=4	10–9=1	10–24=–14	10–30=–20
Xram 4%	**15+0=15**	**15–4=11**	**15–8=7**	**15–12=3**	15–32=–17	15–40=–25

A result here that seems, at least initially, perplexing, but one that will be important for our later discussion is this: hard-nosed capitalists will think that it is in their interest to have some of their capital *not* make a profit. This may sound counterintuitive; why would a capitalist ever forgo profit? The answer, in short, is that by forgoing a profit on their eBay stock, they will make more money overall. In other words, rich capitalists will find it to their economic advantage to have eBay run on a co-op or socialized model. Not seeking to make a profit will be to their overall economic advantage. The reason is obvious; lower transaction fees have opportunity costs that are borne equally by the low- and high-income eBay earners. Lower transaction fees spread the cost and opportunity cost of eBay's capital to the middle- and lower-income earners.

U.$. Inc.

The thought about distributing shares of eBay to all citizens is, of course, only to illustrate a point. After all, US citizens already own something several orders of magnitude greater than eBay. US citizens have a large number of tangible assets under their control. We previously mentioned roads, but the list is far more extensive, for example, parks, bridges, tunnels, nuclear submarines, airports, universities, federally owned lands, and so on.[6]

The intangible assets of the United States include certain institutional arrangements for buyers and sellers, such as an advanced legal and judiciary system that makes the country an attractive place to do business. Just like eBay, the United States offers a market of trust. Indeed, the market of the United States is several orders of magnitude larger than that offered by eBay, so there is every reason to suppose that the intangible market of trust that United States offers to buyers and sellers is worth several orders of magnitude more than that offered by eBay.

Before we ask how these assets should be managed, it is worth confirming that they are indeed owned by "We the People." There is potential confusion given that individual citizens seem to have limited control over these assets. For example, the average US citizen cannot take one of the US navy's attack submarines out for spin or use it for a weekend fishing expedition with 48 beers and some friends. This line of thought sometimes leads to the conclusion that the government owns the submarine and the other assets of the state. Of course, there is no difference on this point with stock ownership. Just because I own a couple of shares of Apple Inc. does not mean that I can go to any Apple store and walk out with an iPhone without paying. Still, it is wrong to conclude that the management of Apple owns the iPhone. It is owned by Apple Inc. Apple management *works for* Apple Inc. Similarly, Apple managers cannot simply take iPhones off company shelves—they too must pay for them. And just because I cannot take an attack submarine out for a spin does not mean the government owns the submarine. The submarine is owned by the people. The government *works for us*: they manage *our assets*.

Here is another way to confirm that we the people own the assets of this country. Suppose we elected a libertarian government that wanted to sell off all the assets of the United States to the highest bidder. As roads, parks, police forces, the court system, the military, and infrastructure are sold off, the government collects lots of cash. If the sitting government tried to vote to give the trillions of dollars collected to the members of Congress, there surely would be riots in the street. The expectation would be that the funds collected would be divided equally among citizens.

It should be obvious that US assets are underutilized in terms of their money-making potential. Indeed, it is hardly an exaggeration to say that the country is run like a giant hippie co-op that consistently loses money. To remedy this situation, we should consider a more thoroughly capitalistic spirit for managing the United States. To solidify this thought, let us imagine dissolving the money losing co-op known as "the US" and forming a corporation, "U.$. Inc.," which would be in charge of running state assets. As with the eBay thought experiment, every citizen of the United States would be given a single share in U.$. Inc. and also one vote at the shareholders' meeting to elect a CEO for U.$. Inc.

Let us suppose our same two candidates reappear. Both agree that it would be illegitimate for U.$. Inc. to try to impose a tax, in the same way that eBay cannot impose a tax. Rather, what U.$. Inc. will do is impose transaction fees for using its market place and all the benefits it provides, just like eBay imposes transactions fees for using its market place. (Alternatively, we can say that eBay imposes a tax on transactions between buyers and sellers; it makes no difference for our argument.) Bush runs on a platform of lowering transaction fees. As with eBay, he argues in particular that lowering transaction fees for the highest earners is important, since they are the ones largely responsible for the glory and prowess of U.$. Inc.

Xram argues that Bush's proposal is ridiculous; U.$. Inc. is a viable business that ought not to operate as if it were some dilapidated hippie co-op. Xram argues that the CEO's primary responsibility is to make a good return for shareholders. Xram promises a dividend of $10,000 per adult citizen. While this may seem an impossibly large dividend to offer shareholders, Xram uses the analysis of the previous chapter to show that it is easily achievable. The 14 percent VAT proposed is rebranded as a "transaction fee." The proposal then, is that U.$. Inc. will have a flat 14 percent transaction fee on all final goods and services.

Xram's proposal, to leverage U.$. Inc.'s money-making potential and distribute the return as a BIG dividend, would be a vast improvement over the contemporary situation that many of the poorest of the poor face: homelessness and hunger. As we noted, BIG provides an income approximately at the poverty line applicable for many places in the United States. In some larger and more expensive cities, it may still be impossible to live a minimally decent life.

Notice, then, that Xram's proposal is not the welfare state theorists' understanding of giving public money to the poor, but capitalism at its finest; making money off of capital. If it sounds ludicrous that people should get money for just lying around doing nothing, consider that there is no work requirement for the super-wealthy who choose to live off their stock dividends. Xram promises everyone just that; a stock dividend for citizens for their ownership in U.$. Inc.

Given the choice between Bush and Xram, it might seem that the obvious thing to do is to vote for Xram. But as with the eBay analogy, it will depend on one's level of economic activity. Again, most people would benefit from a 14 percent transaction fee coupled with a BIG of $11,400. Yes, they would pay more for individual items, since now there would be a transaction fee, but most will have more money in their pockets to begin with. So for the vast majority, running U.$. Inc. like a business that provides a dividend is in their financial interest. Instead of giving away at no cost the wondrous advantages of the US market place, it is in their interest to ask for a return on their investment.

The argument may seem to be faced with this paradoxical result: making money off of state capital is not in the interest of the crème de la crème of the capitalist class, including many of the top income earners. How can this be a proposal for full-time capitalism if it goes against the interest of the capitalist class? The answer is that it is not always to the benefit of the rich for capitalism to hold sway. Consider that in the Great Recession (2007–2009) governments bailed out many large corporations. The rhetoric used to support this was that some companies are "too big to fail." Critics complained that big companies take advantage of this in that they privatize profits and socialize the risk. Specifically, in the good times, private individuals took profits from these companies, and in the bad times, public money was used to bail out these same corporations. Certainly the advantages to the rich of applying capitalism inconsistently are obvious. Instead of losing their capital during the Great Recession, which would have happened if capitalism were consistently applied, many of them benefited from state intervention. The same is true with managing state capital in a socialized fashion. Clearly, it is in the self-interest of the rich for state capital to run on a socialized model and for non-state capital to run on a capitalist model. This is why the rich should vote for Bush if they want to maximize their own economic self-interest. Bush's proposal allows profits to be private when it is to the advantage of the rich and allows cost to be borne socially when it is to the advantage of the rich.

At the risk of belaboring the point, let me use a more homely analogy. Imagine you and nine neighbors own Private Street. You each pay $200 a year for upkeep of the street. At the annual meeting to determine the management of the street for the coming year, one of your neighbors proposes putting a tollbooth at the end of Private Street. It is determined that there are about 100 vehicle trips per day on the road and if only 25 cents were charged per trip, the community would have $25 in revenue per day from the business venture. This works out to $9,125 per year. The upkeep on the road is $2,000 per year, so the net profit would be $7,125. Divided by the ten owners, this would mean a modest profit of $712.50 per year. Someone then points out that this means the average person will spend $2.50 on tolls per day, which works out to $912.50 per person, per year. So this means the tolls will exceed revenues by $200 a day. So, the average person will be back exactly where he or she started.

The reasoning of the previous paragraph is exactly right. On the proposed taxation or transaction fee, the average person will not make anything on Private Street nor on BIG. Both will be perfectly revenue neutral.

Why then should we adopt a transaction fee if the average person will not make anything? The short answer is as obvious as it is painful: *most of us are below average*. While there is an average of ten trips per day, per resident of Private Street, most of the residents make far fewer than ten trips a week. One resident does not even own a car: she walks everywhere in an effort to reduce her environmental footprint. Seven of the other residents make an average of four trips per day. This means that eight of the residents account for 28 trips per week. Of the two remaining residents, one is a lawyer with a home-based business. She has clients coming and going, so she accounts for 12 trips per day. The tenth resident has a home-based business importing and selling electronic accessories. There are delivery trucks constantly dropping off and picking up electronic goods at his home, such that he accounts for 60 trips per day. On Private Street, most residents are below average in terms of their use of the street. One resident, the lawyer, is slightly above average in use (12 trips versus the average of ten), while the electronics entrepreneur has six times the average usage (60 versus ten). The situation is not much different with income distribution in the United States. Consider the difference between the average income, $38,000, and median income, $26,000.[7] The average simply takes the gross income of the population and divides by the number of people. The median income represents the half-way point where half the people in the population make more and half make less. In other words, if you make $26,000, then half of your fellow citizens make more than you and half make less. If you make $38,000, then approximately a third of the population makes more than you and two thirds make less. Average income is so much higher than median income because a small number of people make an extraordinary amount. On Private Street, the median number of trips per day is four while the average is ten.

The bottom line is this: if you are among the top earners, it is in your interest to not make a dividend on your share of state capital. Indeed, you should try to dupe society into thinking that state capital is not the sort of thing that people should make a dividend on. It might help to vilify the government as inefficient and wasteful. If you are part of the oppressed majority, then you should agitate to make a dividend on your share of state capital. You should be opposed to the parasitism of the rich who expect you to support an economic system that permits them to make a profit on their capital but not you on yours. You should oppose the rich because they hope to dupe you into believing that it is natural they should take the benefit of state capital without paying the full cost.

More on the Stockownership Analogy

One objection to the argument is that ownership of stock in U.$. Inc. is very unlike ownership of stock in eBay. After all, one can buy and sell stock in eBay. Suppose, for example, every US citizen was given one share in eBay. We would expect a different pattern of ownership of stock to quickly emerge. We might imagine that many homeless people would sell their single shares for food, drug addicts for drugs, students to pay for student loans, and families without health insurance to raise cash for a lifesaving operation for one of them. So, quickly a pattern would emerge, with most of the shares held by the rich and very few by the poor. We should expect exactly the same thing if U.$. Inc. were a realistic analogy.

I concede immediately that the analogy is not perfect. It might be revised in this manner. Ownership of U.$. Inc. is like a partnership agreement with the following stipulations: every partner's share is equal to that of every other partner (US citizen), and shares in the partnership cannot be bought or sold. Not all partnerships are like this, but it is hardly beyond the realm of imagination to suppose that three lawyers might create a partnership agreement where each has an equal say in how the company should be run and an equal share of the profits. In order to ensure full commitment, it is further stipulated that should a partner decide to leave, her share of the company would be split between the other two. I suggest we think of U.$. Inc. along the lines of such a partnership agreement. Several points can support the analogy, which suggests you can own something even if you cannot sell it.

First, suppose the analogy of the partnership is flawed and we permit the selling of the share of U.$. Inc. much like stock is traded now. If so, then it seems that the selling of one's share of stock in U.$. Inc. would involve the selling of one's vote. In which case, when stock is sold, some will have more votes than others. But this violates the long held belief that it is not permissible to sell our votes, even if it would make perfect economic sense to do so. A penniless person might gladly sell his or her vote for very little, yet we prohibit this. Our vote is ours, we cannot transfer or sell it to another. Of course, we can give up our right to vote, for example, by renouncing our citizenship, but this does not tell against the point that there is an absolute prohibition on selling one's vote.

Perhaps it will be protested that we might permit selling of one's share in U.$. Inc. but not one's vote. That is, we could disengage the political, the right to vote, from the economic, the ownership of the share of U.$. Inc. So, even if shares of U.$. Inc. are sold, one would still retain the right to a say in how U.$. Inc. is run.

To see why this move should be resisted, think about the parallel with slavery: we are all partners in a society that rejects slavery. It was a political decision to outlaw slavery and one that drastically curtailed certain forms of economic bargaining. For example, if the political decision to outlaw slavery were not in force, it might make perfect economic sense for the contracting parties to enter into a master/slave relationship. Imagine a surgeon agrees to perform a lifesaving operation for the mother of three children on the condition that the husband agrees to become the surgeon's personal slave. Suppose too that the family is economically destitute. We can see that, at least from the husband's perspective, this may seem like the most rational option. Without the surgery, his wife will die and his children will grow up motherless. If he agrees to be a slave, his wife will live and his children will have a mother. He knows that the surgeon is not without some compassion and so the husband has good reason to think he will have at least some contact with his family. Indeed, the contract allows the husband to have Sundays off, as long as the surgeon is not golfing. There can be no doubt that if the law did not prohibit it, slavery would reappear (but along class, rather than racial, lines). Drug addicts might happily sign off on a Faustian bargain: two years of carefree drug use at the prospective slave-owner's expense, after which they would become slaves. Nevertheless, one important reason not to allow slavery is that it would tend to drive down the value of U.\$. Inc. Many people find slavery morally reprehensible. It is possible to imagine that many would move away from U.\$. Inc and trade would go down if U.\$. Inc. permitted slavery.

Similarly, if we permit people to sell their share of U.\$. Inc., then it is foreseeable that some people would become financially destitute. For example, suppose you sell your share of U.\$. Inc. to a well-off person for \$100,000.[8] The purchaser now receives your \$833 a month until you die. The rich person takes some risk that you might die early, but hopes to earn a high expected rate of return over the long run. Suppose further that you take the \$100,000 to Vegas and gamble it all on number 36 on the roulette wheel. You are now penniless and without a job. As a society, we must either feed you or let you starve. Both options hurt U.\$. Inc. If we feed you, then our overhead will go up, as we will have to pay for you and others who have given up their share of U.\$. Inc. and have no other means of support. Alternatively, we could simply let you starve, but this will tarnish the image of U.\$. Inc. in much the same way that slavery will. So, it would be bad business to change the partnership agreement and allow people to sell their shares of U.\$. Inc.

There are also mildly paternalistic and social reasons for a dividend from state capital rather than a lump sum. Some people might actually gamble or otherwise foolishly spend their lump sum capital.[9] In any event, whether we allow people to sell their share of state capital or not, this will do nothing to show that our present part-time capitalism is justified. In other words, whether we permit the selling of state capital or not does not change the fact that our system at present is a part-time capitalist one. A part-time system that allows the rich to take the benefit of capitalism without paying the cost.

Socialism

We can see the baselessness of the objection that BIG concedes anything to socialism. As we have said, the point of full-time capitalism is to extract a profit from ownership of all the means of production: the very essence of capitalism. The objection that full-time capitalism does not require work, that the Malibu surfer may enjoy the waves without working, does not make it a form of socialism. After all, as noted more than once, while full-time capitalism permits people to not work and live off of their dividend, so too does part-time capitalism. As is obvious from the pictures of the idle rich that festoon the front covers of tabloid magazines; part-time capitalism permits the wealthy to live off of dividends from their wealth. Full-time capitalism simply broadens the class of capitalists.

Indeed, as intimated, part-time capitalism is more socialist than full-time capitalism. Part-time capitalism aims to administrate the resources of U.$. Inc. as if the aim were *not* to make a profit. As we have just noted, the effect of this—intended or not—is to subsidize the rich. Of course, the rich have very powerful rhetorical strategies to prop up part-time capitalism. They may claim that the poor are lazy and governments are wasteful. From the point of view of full-time capitalism, these criticisms are mere red herrings that mask the truth: *not providing a BIG dividend means that the rich are subsidized by the poor.*

Coercion and Capitalism

An important objection to U.$. Inc. is that there is a major dis-analogy between the profit-seeking activity of U.$. Inc. and its other business activities, namely, that U.$. Inc. is able to extract wealth only because it is a monopoly backed by illegitimate use of force. The thought is that there is choice in the market place: we can choose between different car brands, restaurants, and so on. Indeed, the eBay analogy supports this claim: there are alternative online auction companies to eBay if one does not like the price or services offered by eBay. The state offers no such alternatives and so it is a monopoly. And part-time capitalists have no choice but to comply with the dictates of the state.

Calling the property of the state "U.$. Inc." does not change the fundamental fact that the state is a monopoly.

In response, I will begin with a small but important point: the higher transaction fees proposed by U.$. Inc. are not on wealth, but on income. Wealth tends to be backward looking: it is the accumulation of financial resources that happened in the past. The higher transaction fees are not going to be applied retroactively to income or wealth generated in the past. Rather, the higher transaction fees are to be applied to income in the future.[10] So, a rich part-time capitalist could avoid all future higher transaction fees simply by refusing to generate any income using the means of production owned by U.$. Inc., just as one could stop using eBay if one felt the transaction fees were too high.

Of course, the part-time capitalist will say this, in effect, equals the choice between not having an income or paying exorbitant transactions fees to U.$. Inc. This is to say, not much of a choice at all. But this is exactly the same choice that workers have in a capitalistic society, that is, workers are free to refuse to sell their labor to capitalists or to not have an income. Indeed, many rich capitalists are in a better position: if they refuse to make money, they can live off their wealth, at least for a time. Workers will quickly end up living on the street if they refuse to work, as they typically have no capital to live off.

Furthermore, U.$. Inc. does not hold a monopoly. Part-time capitalists have all sorts of opportunities to not use the services of U.$. Inc. A part-time capitalist could go on the market and look around to see what other companies are offering. A capitalist, for example, could compare U.$. Inc. with its rivals, U.K. Inc., Somalia Inc., and so on. There are more choices in other states for the wealthy to live than there are for many consumer goods, for example, there are only a handful of car manufacturers and cell phone providers from which to choose. In terms of consumer choices, the major corporations in the United States more closely approximate an oligopoly than do the states of the world. There are literally hundreds of countries and so, on our analogy, state companies, from which to choose.

Of course, a part-time capitalist may not like that she must move to get a better price on state services. There should be more local competition. But surely the test is not whether we like the choices. Indeed, one of the surest means to maximize profits is to make the choice of not paying a premium over competitive market prices too high for consumers. Ask yourself this: how can Disney World charge $10 for a hotdog when a street vendor may have trouble getting $2 for a qualitatively identical hotdog? How can a movie theater charge an exorbitant price for a small bag of popcorn? The answer, of course, is that Disney World holds an effective monopoly within its large borders. The cost of going miles outside of those borders to get a competitively priced hotdog is too great for most consumers. Similarly, the movie theater will not permit you to bring popcorn inside its borders and so can charge a premium.

Consider too that individuals and corporations have extremely large landholdings. Disney World is larger than seven internationally recognized countries and territories. Or, consider John Malone, who owns 2,200,000 acres:

> Britain's Daily Mail was flabbergasted—and rightly so. The amount of acreage owned by Liberty Media's Chairman John Malone is beyond the comprehension of city slickers, suburban dwellers, and even plain country folk. As the Mail put it, the 71-year-old's holdings are "just under the size of the Middle Eastern country Jordan and just over the size of Serbia." That means the total sum of Mr. Malone's land is nearly three Rhode Islands. Or two Delawares. It is the size of 151 Manhattan islands. It's a lot of land.[11]

In terms of area, Mr. Malone owns more land than 136 internationally recognized countries and political territories.[12]

Finally, consider what a part-time capitalist will say to complaints about Disney's $10 hotdog: "If you think they are charging too much, then you should think of this as an excellent business opportunity. You should set up your own theme park and undercut Disneyland on the price of hotdogs." The implicit point here, of course, is that when you think about the difficulty and risk in setting up competition to Disneyland, you will see that $10 for a hotdog is a fair market price. But then similar advice can be given by the full-time capitalist to the part-time capitalist: if capitalists think the transaction fees of U.$. Inc. are too high, then this should be seen as a fantastic business opportunity.

Some libertarians, for example, have considered buying an island or setting up a nation at sea, perhaps refurbishing an ocean liner. An appropriate name might be, "Our Lady Of Taxes Are Theft." I predict that should she ever set sail, Our Lady Of Taxes Are Theft will sink—qua business model. For if it is launched as a business venture where it seeks to give a return to investors, it will have trouble competing in the market place for the usual reasons: would-be entrepreneurs who think they can undercut the price of existing business typically underestimate the costs of doing business and overestimate revenues. If Our Lady Of Taxes Are Theft is run on a socialistic co-op model, then I predict it will not work for the reason that many co-ops fail: it is hard to apportion the costs and benefits of such ventures. For example, if everyone living on Our Lady Of Taxes Are Theft has to pay the same amount for upkeep of the ship and its defense, I may find it unfair if my neighbor has 100 times my wealth. After all, the ship will be more of a target for Somalian pirates because of my neighbor's wealth and so incur greater expenditure on defense. I might insist that my neighbor pay more. My neighbor may feel that this is just the progressive sort of tax structure that he sought to escape by moving to Our Lady Of Taxes Are Theft.

But supposing these problems are overcome, will not Our Lady Of Taxes Are Theft be in a much more competitive position than U.$. Inc.? After all, the lower transaction fees of Our Lady Of Taxes Are Theft should make it an attractive place to do business.

There are at least two major problems with this suggestion. First, if all there is to making a business decision is looking for the lowest price, then one should predict that eBay would be one of the smallest online markets. After all, it charges more than most other online auctions. However, people are willing to pay a premium to be part of a much larger market. Perhaps it may be thought that capitalists on Our Lady Of Taxes Are Theft will be able to use the market of U.$. Inc. but not have to pay the transaction fees because they live at sea. This same brilliant strategy could be used to get around the transaction fees of eBay. Some sellers, for example, imagine that they might advertise their wares on eBay but sell them privately in a way that avoids the eBay transaction fees. eBay works hard to stop such attempts at parasitism, and so should U.$. Inc.[13]

The larger of the two major problems is how Our Lady Of Taxes Are Theft plans to attract workers for capitalists to exert their vampire-like effect. For example, the promise of lower transaction fees on Our Lady Of Taxes Are Theft is not going to attract low-wage earners. A minimum wage earner in U.$. Inc. might say to himself, "I earn $13,000 a year here, and pay $2,600 in taxes. The flat tax of Our Lady Of Taxes Are Theft is half that, so I would have to pay only $1,300 in taxes. But here in U.$. Inc., I also get my stock dividend of $11,400. So even with the higher income tax rate and VAT, I am still farther ahead here with my combined income (wage and BIG) as compared with the possibility of earning $11,700 on Our Lady Of Taxes Are Theft. Such a move would be stupid simply in terms of earnings. Moreover, if I lose my job on Our Lady Of Taxes Are Theft, I have no other economic security; I will be thrown overboard." Of course, capitalists on Our Lady Of Taxes Are Theft could offer a higher wage and unemployment insurance, but then there are serious reasons to doubt the claim that Our Lady Of Taxes Are Theft has some competitive advantage.

Suppose the part-time capitalist complains that the reason Our Lady Of Taxes Are Theft will fail is because the best spots on earth for setting up a competitor to U.$. Inc. have already been taken. I think this is a very good complaint, but unfortunately for them, it is not one that the part-time capitalist can make. The trouble, again, is that workers have at least as strong a complaint against capitalists. Workers who would like to set up an alternative to part-time capitalism are even more out of luck. Typically, they cannot afford even to buy a used cruise ship like Our Lady Of Taxes Are Theft to erect their socialist paradise. Workers have only two options: violent revolution or political change within the existing political structures. Rich capitalists who hope to keep part-time capitalism have these two options plus the option to finance their own state. So again, for every complaint that the part-time capitalist has against the full-time capitalist, workers have at least as strong a complaint against part-time capitalists.

Societal Ownership

It is worth pausing to consider the argument of this chapter in terms of previous thinking. As noted above, the idea of a basic income has a long history. In most general terms, the idea of justifying BIG has had to work two desiderata. One is that sufficient income or capital must be found to finance BIG. The other is to find morally compelling reasons to underwrite the means to finance BIG. These two desiderata are clearly related but often pull in different directions. For example, many think that it is morally permissible for governments to tax natural resources and use the taxes to finance public interests.[14] Indeed, the Alaska Permanent Fund is one example of this thinking in action. Citizens of Alaska typically receive about $1,000 from the fund each year as a dividend from Alaska's oil. Although this is certainly a start, $1,000 is far short of even the most modest of basic incomes. The thought then would be to expand the scope of natural resources to be taxed. Certainly, taxing mining operations seems like a natural analog to taxing oil, but again the problem is that this will not generate sufficient income to pay for BIG. One could imagine taxing the consumption of air, and certainly people would pay a large amount for air under threat that it would be taken away, but the problem is to find a morally compelling reason to tax people for the consumption of air. At least in the case of oil and mining, there is an obvious rationale for taxing: the supply of oil and minerals is limited, and no one made these resources. It is true that no one made the atmosphere, but the supply of air is not limited (in any relevant sense for political philosophy).

A more promising way to finance BIG is suggested by various "cost of civilization" arguments. The basic idea, going back to Thomas Paine and recently championed by Karl Widerquist, suggests that our current capitalist system imposes costs.[15] For example, a system of private property and capitalist market exchanges makes it impossible for almost everyone in the United States to live a life like most humans lived throughout the course of human history: the life of a hunter gatherer. Justice requires that we compensate people for this loss. We must, to use Widerquist's apt phrase, provide people with an "exit option."[16] As Widerquist argues, a minimally acceptable exit option is a basic income to support people who would rather not participate in the current economic system.[17] Given the general premise that we have a duty to compensate people for the cost of civilization, it is an easy run to the conclusion that we ought to finance BIG through taxation. The trouble is that the premise that we ought to compensate people so that they may opt out is controversial from the point of view of contemporary capitalism. What I mean by this can be illustrated with an example. Imagine the following response by Joe Average citizen, "I bought a house out in the country to get away from the noise and the pollution of society. Well, just my luck: a big developer put in a shopping mall and thousands of new houses in a new development right next to me. You don't see me whining to the government about payment so I can have an exit option from living in suburbia. So why should I pay more taxes so that others can drop out of the system?" I'm not saying that we should find this rebuttal convincing. I am saying, however, that it puts a lot of weight on the moral principle that we ought to compensate, one that capitalists will need to be persuaded of.

A similar problem faces van Parijs's defense of BIG. Van Parijs considers taxing inheritance, natural resources, and technology but concedes that this will not generate sufficient income to finance BIG.[18] To finance BIG, then, van Parijs proposes to tax high income earners on the moral basis that high-paying jobs are a scarce resource. This argument has come under trenchant criticism,[19] and even those sympathetic have suggested the argument must be reworked in significant ways.[20] Assuming the argument can be put in satisfactory form, it still relies on a premise about redistribution of resources. Imagine the same capitalist responding to van Parijs, "My brother made at least 10 times what I made before we retired. He will be the first to tell you that I have more business smarts, more drive, and more people skills than he does. He just happened to be at the right place at the right time: I was not so lucky. But you don't see me asking the government to tax him to support me. There is some brute luck involved in business and so in capitalism: you pay your money and you take your chances." I'm not saying that we should find this rebuttal convincing. I am saying, however, that it puts a lot of weight on the moral principle that jobs are resources that should be taxed, one that capitalists will need to be persuaded of.

The present argument does a better job of satisfying the two desiderata because it meets the capitalist on his home turf. In terms of financing BIG, the class of resources to be leveraged in the service of BIG is much larger than is typically proposed. The transaction fees are on all economic activity, which includes all the aforementioned items to be taxed: the sale of oil and other natural resources, land, and high-paying jobs. Of course, the proposal here would add a transaction fee on many additional items. The moral premise of the argument is one that the capitalist will find harder to resist: the key claim is that it is permissible for us to leverage our assets to try to make as much profit as possible to serve our own interests. Indeed, one may think this is the very essence of capitalism.

Conclusion

The poor in this country must come to realize that they are rich: they have a share in potentially the greatest money making machine in history: U.$. Inc. They should demand of their CEO (President) and board of directors (Congress) a good return on their investment.

We should give a nod to those who did the lion's share to create state capital: our dead ancestors. Those who fought in wars, those who fought for justice, and those who played by the rules are just a few examples of the sorts of activities that contributed to making the United States a rich market place. We are all a little like Paris Hilton: lucky inheritors.

I am not saying part-time capitalists will like this. We have shown how making a profit on the state's means of production would be to the financial detriment of the wealthiest. As we have said, part-time capitalism works wonders for the rich. Nor am I saying it will be easy to convince rich capitalists. As Marx noted, capitalists are as much subject to false consciousness as the proletariats. What I am saying is that the internal logic of their view of distributive justice means that something much like BIG could be justly instituted, whether the rich believe it or not. As noted, it is in their interest to propagate the ideology that transaction fees are theft. But this is just to say that it is in their interest to propagate an ideology to hide the truth. And while they may continue to do so with a disproportionate sway in public discussion, we can take solace in the fact that, at election time, the numbers favor those who would benefit from full-time capitalism.

Furthermore, I am not saying that socialists should be appeased by a BIG dividend. Full-time capitalism would permit large gaps in economic equality, which socialists would see as unjustified. The severely mentally or physically disabled, for example, might not be able to manage on $10,000 per year or perhaps even more if the market will bear a higher transaction fee, yet full-time capitalism does not address this question. In other words, a basic income is a floor-level income rather than a ceiling-level. If one thinks that distributive justice ought to take into account the special needs of some citizens, then the floor level for some may be raised much higher.

What I am saying is that part-time capitalists should have no legitimate complaint should the oppressed economic classes vote themselves a BIG dividend. The upshot of the inverted Marxism argument is that every complaint by part-time capitalists that this proposal is unjust can be met with an equally strong complaint by workers against part-time capitalism.

CHAPTER 4

Capitalism: Consequentialism versus Rights

Introduction

In this chapter, we will be exploring consequentialism as a means to defend BIG. Consequentialism says that the correct scheme for property distribution is the one that maximizes the social good. Since taxes tend to promote the social good, taxes are justified. Individual ownership of property must be tempered in service of the overall good of society. Some proponents of rights-based political theory reject the idea that property rights might be limited to the service of overall social good. That is, they reject consequentialism as a basis for forming sound public policy. We will examine the rejection of consequentialism in this chapter. In particular, our concern is whether the idea of property rights can be used to defend the claim of capitalists that taxes are theft. We will look first at the rallying cry that taxes are theft.

Taxes are Theft

The phrase "taxes are theft" is familiar to anyone with even a passing acquaintance with American politics. One might think the phrase is uttered only by antigovernment loons holed up in bomb shelters with assault rifles. However, the phrase has found a very prominent spokesman in Speaker of the House John Boehner, who asked, "How much more money do we want to steal from the American people to fund more government?"[1] Boehner's answer is, "I'm for no more." If taxes really are a matter of stealing, then it seems that Boehner has not objected strongly enough. Of course, we should not want *more* stealing, but that is simply because we should not want *any* stealing. (To the question, "How much more rape do we want in this country?" presumably saying "I am for no more" does not quite get to the heart of the matter.)

The appeal of the "taxes are theft" slogan is as seductive as it is simple-minded. It often starts with a thought along the lines embodied in this sort of example:

> **The tomato farmer: Taxes are Theft:** I grew some beautiful succulent tomatoes through hard effort—and my effort alone. The government did not help me water them every day. The government did not help me turn the soil, or fend off the aphids. I am about to take them to the market where they will fetch a good price. And this is where the government bandits will intercede: they will take a cut of my profits and call it "taxes." This is no different than if a band of highway robbers took some of my profits at the end of a gun on the way home from the market. In either case, some of my sweat and toil is taken from me.

This example may illustrate the thinking that taxes are theft, but it is not going to persuade the unconverted. Indeed, for those who believe that taxes are a legitimate part of the economic system, the argument simply begs the question.

Consider that the description of the tomato transaction assumes that it is a two-party transaction: the tomato grower and the tomato consumer. On this understanding, any third-party claim is illegitimate. The assumption makes for an easy run to the conclusion that any government involvement in the transaction is theft. By the same thinking, the tomato grower could show that eBay's demand for 3 percent of the transaction, if the tomatoes were listed on eBay, is also theft. For those who think the buying and selling of tomatoes is actually a three-party transaction—the buyer, the seller, and the government—then failure to pay taxes is theft from one of the parties, namely the government. What counts as "theft" is intimately related to assumptions about economic legitimacy.

The point can be made more perspicuous with the following example: a southern racist claims that the actor, Morgan Freeman, has been stealing his money. He bases his claim on the fact that his family owned Freeman's family before the government stole his ancestors' slaves with Lincoln's Emancipation Proclamation. Since a slave's children were owned by the slave owner, and all labor of a slave belongs to the slave owner, Morgan Freeman's acting fees and all his property belong to the racist. The racist accuses the government of supporting Morgan Freeman's theft.

Of course, this argument turns on the assumption that the economic system involving slave ownership is the correct one to apply to the question of the ownership of Morgan Freeman's bank account and other assets. If this assumption is correct (and clearly it is not), then the racist would have a point about theft. Since we reject this assumption, we can reject the claim of theft. Or, to put the point the other way around, if the racist wants to make the charge of theft stick, he would have to first convince us that ownership of slaves is legitimate. Otherwise, the claim that Morgan Freeman and the government have stolen from him simply begs the question.

Similarly, Speaker Boehner and likeminded fellows need to establish first that it is illegitimate for the government to impose taxes. If they can establish this, then the charge of theft may have some substance. But until they establish this, calling taxes theft simply begs the question at hand. In other words, to simply assert taxes are theft is to argue *from* the assumption that taxes are not a legitimate part of the economic system. But for those who disagree with this assumption, it simply begs the question. What Boehner and others need to do is to argue *to* the claim that taxes are not legitimate.

Of course, the point cuts both ways, those who charge that tax evaders as stealing from the government also require an argument for their claim. To simply assume that taxes are legitimate is to argue *from* rather than *to* a distribution scheme that includes taxes. So, we need to probe further into what might justify an economic system of distribution.

Economic Systems and Economic Policy

Before we look at arguments for different economic systems, it will help to discuss about economic systems in general. An "economic system" encompasses all the different facets of the economy of a country, society, or community. Every economic system has rules or norms governing the production, consumption, and distribution of economic goods. Some important historical examples of economic systems include slave societies, feudalism, state socialism, and capitalism.

Of course, there are more economic systems than the few just mentioned. Marx, for example, in his historical studies of economic systems, mentions "primitive" forms of communism: the economic system of our distant ancestors who lived in small tribes (perhaps no more than a few hundred people) and shared their economic goods among tribe members. Also, the economic systems mentioned can overlap (they are not mutually exclusive). The slave society of nineteenth-century America had many capitalistic elements. Slaves, for example, could be bought and sold at the open market. Today, employers sometimes require noncompetition contracts from their employees; for example, if employees leave the job, they cannot go to work for a competitor firm (nor start their own competing firm). To some, at least, this sounds vaguely similar to the lord/vassal arrangement where the vassal is tied to some means of production via contractual arrangement.

By "economic policy," we mean the societal norms that govern economic activity. A radical change in economic policy can result in a change from one economic system to another. Outlawing slavery or lord/bondsman relationships resulted in a change in economic systems. Economic policy changes obviously need not be so radical. The law of 1808 banning the importation of slaves into the United States resulted in a change *to* slave society, but not a change *from* slave society. Since the slave trade treated slaves as mere things, the trade responded as one might expect when there is commodity shortage from abroad: it increased the domestic production of slaves. So by the locution "economic policy decisions," we mean all decisions about the economic workings of a society, from large-scale decisions about which type of economic system to adopt, to small scale changes within a system. Accordingly, the decision whether to allow shopping past nine o'clock in the three-block downtown area in Small Town, United States is as much an economic policy decision as the adoption by Russia of communism as its economic system in 1917, or as Lincoln's Thirteenth Amendment outlawing slavery.

We Should Judge Economic Policy in terms of Aggregate Well-being

The proposal of this work is to evaluate economic policy and economic systems in terms of how well they promote aggregate well-being. By "well-being" I mean how well a life is going for the person who is living it. An answer I will defend in subsequent chapters is that happiness and freedom are two important components of well-being. Aggregate well-being is the sum of individual well-being. To decide between different policy options requires assessing how each policy would affect the happiness and freedom of all those involved. The policy that will result in the highest net happiness and freedom for the society as a whole is the one that should be adopted.

To work a simple example: suppose we are asked to decide between our present economic arrangements and returning to racially based slavery. To compare these alternatives, we would look to see which system provides more overall happiness and freedom. Clearly the race chosen to be slaves would be much unhappier on this proposal and much less free. Still, this does not decide the issue. After all, if we are looking at aggregate happiness and freedom, we would have to consider the possible consequences for the well-being and freedom for the rest of the population. For the rest of the population too, it seems likely that both happiness and freedom would decrease. Think of the howls of protest from the vast majority who believe that slavery and racism are wrong. Many, not just the newly enslaved population, would be deeply saddened by such a morally depraved economic system. Perhaps a few racists might be happy about such a policy, but their happiness would be entirely swamped by the unhappiness of the rest. So, judging in terms of aggregate happiness and freedom, racially based slavery today would be a very bad policy in comparison to present arrangements. We can see then how "consequentialism" gets its name. Economic policy is evaluated in terms of its consequences for aggregate well-being: the best policy is the one that does the most to promote aggregate well-being.

We will argue on the matter in more detail shortly, but it should be apparent how consequentialists might see enormous benefit in taxation. Empirical research indicates that the well-being of the rich is only marginally affected by high taxes, but when taxes are distributed, it can have enormous positive effect on the well-being of the poor. At this stage, we can at least see this much: there is no principled reason for or against taxation, according to the consequentialist. Whether taxes are a good idea or not will depend on their effect on the well-being of society as a whole.

Taxes are Theft: Rights Reboot

One thing consequentialism seems to say is that it is permissible to sacrifice the well-being of individuals for the greater good. Introductory philosophy classes often include thought experiments about whether it is permissible to kill one person to save five. Consequentialists will often sanction the killing of one person to save five because this will do more to promote aggregate well-being. Yes, it is a tragedy that one person dies, and this counts as a negative in consequentialist thinking, but the gain in well-being by sparing the lives of five others more than compensates for the loss. Diametrically opposed to consequentialism are deontological theories that deny our right to judge the rightness or wrongness of actions simply in terms of results or consequences. The most familiar forms of deontology are certain rights-based understandings of our moral obligation. For example, if we assume that people have an inviolable right to life, then it would be wrong to sacrifice one person for the sake of five other lives.

This understanding of our moral obligations suggests a different means to defend against taxing for the greater social good:

> Even if taxes will lead to greater social good, taxing my earnings violates my fundamental right to property and so amounts to theft in order to achieve this greater social good.

In the hands of Robert Nozick, the objection about treating some individuals as fodder for a greater good is put this way, "there is no moral outweighing of one of our lives by others so as to lead to a greater overall *social* good. There is no justified sacrifice of some of us for others."[2] Nozick claims that taxes distributed for the benefit of the less well-off are prime examples of unjustified sacrifices.

This argument is clearly an improvement over the bare "taxes are theft" argument. Although the conclusion is similar, that taxes are theft, now we are offered an *independent* reason for thinking that taxes are theft: we have a fundamental right to private property, and taxes violate this right.

We will think of this as the "Lockean argument against redistribution." It has two fundamental premises:

> **Premise 1:** We ought not to violate a person's rights.
>
> **Premise 2:** If we redistribute wealth (as in the BIG proposal), then we violate a person's rights.
>
> **Conclusion:** We ought not to redistribute wealth.

We will look at Premise 1 later, and now discuss Premise 2 in connection with Nozick's version of the Lockean argument that people are "self-owners."

The argument starts with a seemingly innocuous premise: we own our own bodies, talents, personalities, abilities, and labor. If the idea of self-ownership seems a little strange, an implication of it may seem more familiar: if we own ourselves and our labor, then it follows that no one else can legitimately own us—at least not without our consent. Anyone who claims another as property violates this fundamental right. Thus, the right to self-ownership provides a straightforward reason for thinking that slavery is wrong.

Nozick argues that this has implications for the tax system in a welfare state. A tax on income to provide relief for the poor or for other social goods is in effect forced labor: it requires the handing over of money, which is often equivalent to giving over some of one's labor. Since the paying of taxes is not voluntary but coerced, it follows, says Nozick, this is forced labor or slavery (of a limited sort).

I hope to show that the argument only appears persuasive as an apology for laissez-faire capitalism because it turns on an equivocation of the term "rights." In order for us to accept the second premise, the term "rights" must be understood in a very strong way, in terms of what I shall call "Lockean rights." But once we see how strong rights must be understood in the second premise, we can see that the first premise is not particularly plausible. It will take us a while to work our way to this conclusion. We need first to understand the nature of Lockean rights and the connection between Lockean rights and capitalism.

Capitalism and Rights

Let us look first at the claim that the Lockean argument actually constitutes a defense of capitalism. We noted above that capitalism is often conceived as the idea of private ownership of the means of production along with competitive markets. How are these two ideas connected? Perhaps somewhat surprisingly, in the Lockean tradition there is no direct connection between the right to property and competitive markets. Rather, the assumption seems to be that free markets naturally result between individuals who faithfully observe Lockean rights. Without governments interfering to enforce prices, production, consumption, and distribution, free markets naturally develop.

It will help to contrast this with what a direct connection between the two might look like. For example, one way to directly connect the two ideas would be to suggest that we have a right to a free market. However, Locke and Nozick did not advocate for a right to a free market, this would seem incongruous with their basic assumption about the right to liberty and private property. Suppose Locke and Marx planned to go camping with six others.[3] Everyone but Locke agrees to share everything they bring and all the fruits of their labor in common while on the camping trip. If someone catches fish during the trip, the proposal is that the fish will belong to the group, to be divided equally among them. Another will do the dishes for everyone. For the duration of their trip, they propose to abide by the modified Marxian slogan: from each according to his camping ability, to each according to his camping need. Locke proposes that they should not pool their property in this fashion. They ought to have a competitive market to increase the efficiency of their camping trip. Suppose the rest are unmoved by Locke's argument and propose to continue their camping trip in the spirit of camaraderie rather than laissez-faire capitalism. They argue that this camping competition will only induce animosity among the campers rather than encourage a relaxing and fun vacation. On Lockean principles, the others could not force Locke to join the camping-collective: this would violate his right to self-ownership. So, if the other seven campers compel Locke to contribute his fishing bounty to the group, then his rights will have been violated. But, conversely, Locke cannot insist that the seven join him in a camping competitive free market. For example, Locke could not insist on a right to a competitive market because this would force others to renounce their right to self-ownership and freedom to dispose of their property as they see fit. In other words, if Locke insists on a right to a competitive market, this would force others to not adopt other models for economic distribution. This would limit the liberty and property rights of others. If Locke attempted to force the other campers to compete with one another, that would violate their property rights.

So, either there is no right to a competitive market, or the right to liberty and private property is in conflict with, or limited by, the right to a competitive market. As noted, neither Locke nor Nozick assert a right to competitive markets, and we can see that they do not do so for good reason.

The point here is small but crucial: if each person has a right to his labor and the right to dispose of his or her property as deemed fit, then there cannot be a direct connection between rights and free markets. For then this unlimited property right would become limited: it would limit how one could economically cooperate with others, for example, it would forbid socialism between free individuals. The primitive communism that Marx thought characterized many of the small tribes of our ancestors would be illegitimate. Defenders of laissez-faire capitalism will note that the conditions that make the socialist camping trip workable—a few friends going for a short vacation—are not reproduced in large modern societies. There is little worry that spontaneous agreements to form a socialist utopia will develop between complete strangers. Without a government to coerce socialism, there is little worry that socialism will form through individuals agreeing to such a system. Where there is no government enforcement, competitive markets will naturally develop between free individuals. While we will examine this line of reasoning below, it is enough at this point simply to see why Lockeans should be averse to appealing to the right to a free market.

Capitalism and Consequentialism

Consequentialist defenses of capitalism are almost the mirror image of rights-based defenses. With rights-based defense, the right to private property is primary, and competitive markets are merely a consequence of this right. With consequentialism, it is the competitive market that is seen as primary, and the right to private property merely an instrument for realizing the power of competitive markets. In other words, consequentialist defenses of capitalism emphasize the enormous benefits that capitalism brings *to society*. Consider the following quote, praising the productive capacity of capitalism:

> The bourgeoisie, during its rule of scarce one hundred years, has created more massive and more colossal productive forces than have all preceding generations together. Subjection of Nature's forces to man, machinery, application of chemistry to industry and agriculture, steam-navigation, railways, electric telegraphs, clearing of whole continents for cultivation, canalisation of rivers, whole populations conjured out of the ground— what earlier century had even a presentiment that such productive forces slumbered in the lap of social labour.[4]

The quote is from Marx and he was one of the most vehement critics of capitalism. Still, it is interesting to note that he was not blind to the ability of capitalism to produce enormous economic wealth.

Of course, Marx is not alone here. In the century prior to Marx, Adam Smith made much of the incredible potential of capitalism to increase production and the social good:

> Among the savage nations of hunters and fishers, every individual who is able to work is more or less employed in useful labour, and endeavours to provide, as well as he can, the necessaries and conveniencies of life, for himself, or such of his family or tribe as are either too old, or too young, or too infirm to go a hunting and fishing. Such nations, however, are so miserably poor, that from mere want, they are frequently reduced, or, at least, think themselves reduced, to the necessity sometimes of directly destroying, and sometimes of abandoning their infants, their old people, and those afflicted with lingering diseases,

> to perish with hunger, or to be devoured by wild beasts. Among civilized and thriving nations, on the contrary, though a great number of people do not labour at all, many of whom consume the produce of ten times, frequently of a hundred times more labour than the greater part of those who work; yet the produce of the whole labour of the society is so great, that all are often abundantly supplied, and a workman, even of the lowest and poorest order, if he is frugal and industrious, may enjoy a greater share of the necessaries and conveniencies of life than it is possible for any savage to acquire.[5]

According to Smith, one great difference between the "savage" nations and "civilized" nations is the free market, which encourages ever increasing efficiency.

Not only is the consequentialist defense of laissez-faire capitalism the mirror image of the rights-based defense, but so too are its weaknesses. Whereas the rights defense makes clear why private ownership of the means of production is justified, its defense of competitive markets is only indirect. While consequentialist competitive markets are justified directly in terms of their promotion of the aggregate good for society, private ownership of the means of production at best must be indirectly justified. Consequentialists typically take private ownership as a good instrument for building and maintaining competitive markets. On motivation, Smith famously wrote, "It is not from the benevolence of the butcher, the brewer, or the baker, that we can expect our dinner, but from their regard to their own interest." We expect them to be motivated by personal profit. These individual self-interested acts, argues Smith, lead to the greater good of society. As Smith claims: "By directing that industry in such a manner as its produce may be of greatest value, he intends only his own gain, and he is in this, as in many other cases, led by an invisible hand to promote an end which was no part of his intention."[6]

Smith's consequentialist defense of private motivation suggests good consequentialist reasons for not permitting unlimited property rights. The easiest way to see this is in the case where an individual or group has a monopoly. Suppose a town has three transportation companies: a toll highway, a railway, and an airline that provide service to a remote mountain location. They compete hard for customers for many years. Finally, the three owners sit down and sign an agreement setting prices for all three services, much higher than the existing rates. Each company will make a lot more and will not have to worry about price competition. A consequentialist would see this as a bad result overall, since it would limit competition and drive up prices for the consumer. A consequentialist would have no problem using government power to break up such a cartel in the interest of the good of society, even if the business owners howl that this violated their property rights. The consequentialist will be unmoved by this protest since property rights are recognized only to the extent that they tend to promote the good of society as a whole.

Rights, Competitive Markets, and Monopolies

We noted that the Lockean apologist for laissez-faire capitalism seems to assume that capitalism, and hence competitive markets, will naturally evolve where Lockean rights are observed. Why should we believe this?

Often a distinction is drawn between government (or de jure) monopolies and natural (or de facto) monopolies. It is clear why government monopolies cannot develop where Lockean rights are protected: enforcing a government monopoly would restrict one's ability to dispose of one's property as one wishes. For example, if a local town council gave a gas station operator exclusive rights to own and operate gas stations, this would be a violation of the rights of others to do with their property as they wish, for example, to use their land to set up a competing gas station. Natural monopolies, on the other hand, are understood as single providers of some good or service where there are no legal barriers to entering into competition with the single provider. The "single provider" notion includes cartels: explicit agreements between companies to act like a single firm monopoly. For our purposes, it does not matter whether competition is limited because there is only a single company provider or a group of companies acting as a single provider, as in the case of cartels. In either instance, there is no competitive market for the product or service.

Since, as we have just seen, government-sanctioned monopolies violate Lockean rights, our question may be made more precise: why should we assume that competitive markets will always develop in the absence of government regulation of private property?

One answer is this: if you own the only gas station in town and charge extremely high prices for gas, others will see this as an opportunity to go into competition with you. If you own the only railway line into town and charge hefty prices, then others might start competing railways or other services, such as bus and trucking lines. As Anne-Robert-Jacques Turgot observed in the eighteenth century, industries that have much higher rates of return on investment (other things being equal) will attract others into the industry.

> As soon as the profits resulting from an employment of money, whatever it may be, increase or diminish, capitals turn in that direction or withdraw from other employments, or withdraw and turn towards other employments, and this necessarily alters in each of these employments, the relation between the capital and the annual product.[7]

It will help to designate this thinking as the "iron law" of competition: no natural monopolies survive for long. The thinking is this: any natural monopoly that makes higher than average profits for its investors will naturally attract other investors to this industry. When other investors arrive, competition will ensue. So, just as iron filings are naturally attracted to magnets, so too is capital attracted to higher profit industries, such as industries run by monopolies. As more capital is attracted, competition increases. The iron law perhaps is what supports the thoughts of Locke and his heirs that competitive markets will naturally develop.

There is a small complication in stating the iron law in this fashion. Sometimes monopolies are defined as "single providers." Strictly speaking, the iron law would not apply to a single provider who does not try to raise their prices in order to make above-average profits. This sort of situation is safely ignored in most discussions, since the assumption is that individuals will aim to maximize profits. Accordingly, if individuals own natural monopolies, the assumption is that they will raise prices to maximize profits. There is obviously much less room for complaint against a natural monopoly owner who takes average, or below average, profits.

The Water Hole

Nozick himself considers the case of a natural monopoly: someone owning the only water hole in a desert. This person could charge monopoly rates to everyone else dependent on the waterhole. On the face of it, this looks like a decisive objection to the Lockean position since it seems grossly unfair for a single individual to lay claim to nature's bounty and to hold all others hostage to monopoly prices for water. The owner of the waterhole could enforce more draconian conditions on others than any tax imagined by the US government.

Nozick has an answer to this sort of objection, once again inspired by Locke. To see his reasoning, we need to look a little bit deeper into Nozick's theory of distributive justice. Stripped to its essentials, the theory has these features: a theory of initial acquisition, a theory of transfer of holdings, and a theory of rectification of holdings. Initial acquisition is subject to the "Lockean Proviso." According to Locke, we can only take from the natural world up to the point where there is "enough and as good" left for everyone else. If you have legitimately appropriate natural resources and you fashion them into something more valuable, then you are entitled to these holdings because they are a result of just acquisition and your labor, which you own. All voluntary exchanges are legitimate.[8] So a voluntary exchange with others might result in an increase in everyone's holdings, given for example, specialization of labor.

The answer to the alleged water hole counterexample is that such a monopoly is illegitimate because it violates the Lockean Proviso: specifically, whoever claims ownership of the water hole has violated the Lockean Proviso because they have not left "enough and as good" for everyone else. So Nozick could agree the claim of sole property rights to a water hole is illegitimate.

The Water Hole Strikes Back

You, along with 99 others, found a community in a desert around the only water hole for hundreds of miles. You suggest that the water hole should be managed for the good of the community. Water could be sold to pay for community services. Your colleagues disagree: as good Lockeans, they reject outright your proposal as "socialistic" and "utilitarian social engineering." You acquiesce. The water hole is divided equally into 100 shares. Everyone agrees this is fair. The Lockean Proviso is satisfied.

The waterhole soon becomes too polluted to safely drink from it. You set up a purification plant and sell bottled water for $30 a month. No one else is tempted to go into competition with you because you make a slim profit of $1 per customer per month. In other words, your cost of production is $29 per month to run the purification plant and bottle the water. You send out two promotions to your customers. Your first promotion offers customers $30 a month for the rights to their 1/100 portion of the waterhole. The offer is valid only for the first 50 customers. The second promotion offers customers $29.99 a month for the 1/100 portion of the waterhole and the offer is valid only for the first 50 customers. Your customers naively do not consider the possibility of a monopoly being formed because of the 50 customer limit. You start charging $2,000 a day for bottled water. Your customers, of course, complain. They make $20 an hour on average, so they cannot afford such an extravagant price. You offer to pay them $100 an hour to come work for you. They say that they cannot possibly work 20 hours a day. You offer to loan them up to $800 a day so long as they work for you for 12 hours a day. Thus the whole town is put in your service working 12 hours a day and going deeper into debt each day. All this happens according to good Lockean principles. A just acquisition was observed because everyone got their fair share of the natural resource, such that it left "enough and good" for everyone else. Indeed, we went a step further than Nozick and even stipulated that everyone from whom shares were purchased agreed on the distribution. There was justice in the transfer because the exchange between you and the other 99 citizens was freely agreed upon. The result ultimately is far more "slavery" for everyone else in the town than the 20 percent (effective) income tax rate that the rich in this country complain about.

The argument, then, is that even by Lockean premises, it seems that a natural monopoly over a resource might develop. I do not want to press too hard on this argument as this example depends on the monopoly over a natural resource. The Lockean proviso leaves much obscured on this issue. For example, one of the natural resources that should be left in common is land. This means that everyone should be entitled to a parcel of the land of this earth, some of its oil, minerals, and so on. How then can the bourgeoisie claim to own the best parts of the world already?[9] Honoring the Lockean proviso would require a dramatic shift in how society is organized. We will work on an example in the next section that skirts the difficult issues raised by the Lockean proviso.

The Small Town

Consider again Turgot's statement:

> As soon as the profits resulting from an employment of money, whatever it may be, increase or diminish, capitals turn in that direction or withdraw from other employments, and this necessarily alters in each of these employments, the relation between the capital and the annual product.

It is simply false that capital flows in the direction of profits. If anything is true, it flows in the direction of *expected* profits. If you see someone making a fortune one night selling Super Bowl t-shirts for your hometown team, Turgot's statement suggests you should set up a competitor t-shirt stand. But it would not make sense to invest in a t-shirt venture if the team loses that night. The price of t-shirts for the losing side, you should predict, will be extremely low—about the same price as that of a rag. Smart capital flows to where it expects the market to be, not necessarily where the market is.

Consider again the case of the small town with a cartel comprising the toll highway company, the railway, and the airline. Just because there are high profits in the transportation industry does not necessarily mean it is a place to expect high profits to flow from the investment of capital. Suppose you are trying to convince investors to put their money into a new highway. At a meeting in a big fancy boardroom, you point out that the cartel charges $2 a trip and 10,000 trips are made per month. The cost of a new highway will work out to $10,000 a month, so you could charge $1.50 a month and take all the business from the cartel and then some. You estimate the number of car trips will increase to 12,000 per month. So your business will make $8,000 a month. Your investors object that it is naive to suppose the cartel will not change their price. If the cartel charges $1.50 to match, and customers are divided equally between the two companies, then each highway will lose $1,000 a month. The cartel has deeper pockets and so could sustain the loss long enough to drive the new highway project out of business. And even if the investors could find the capital to stay in the fight, the long-term prospects are for merely an average rate of return on investment. For if the new highway succeeds, then cutthroat competition will be restored.

Suppose, following your pitch, another would-be entrepreneur comes in to pitch the idea to the investors of creating a company to buy the toll highway, railway, and airline in a neighboring town. The companies at present are in the midst of cutthroat competition: each company is losing money fast. The proposal is to buy up the companies cheap and then raise the prices for all three. Which proposal will investors go for? Clearly, other things being equal, investing to start up a monopoly in a neighboring town is the best bet. Economic rationality that demands the greatest return on investment requires that investors should preferentially invest in monopolies and cartels and only invest in competitive markets as a last resort.

It may be objected that a cartel comprising the highway, railway, and airline would never be allowed to develop, so the example is fictitious. Governments would act against such monopolies or cartels because they would put a stranglehold on consumers and every other business, since everyone depends on transportation. However, this would be to miss the point entirely. It is true that in most jurisdictions there is legislation against cartels and monopolies, so this example is fictitious in that sense. But recall that the idea under consideration is that cartels or monopolies would never result or quickly dissolve because of the iron law. However, this runs counter to the thought that investors will look to invest where there is the best expected return on their investment. When there is a chance to invest in cartels or monopolies, we should predict a massive flow of capital toward such opportunities. Thus, there is every reason to suppose that natural monopolies will develop so long as there is no legislation to prohibit such activity. There is little reason to suppose that natural monopolies will not develop if they are legally permitted.

History

History does not bear out the prediction of the iron law that natural monopolies will not survive long. There are a number of examples in the history of economics regarding natural monopolies or near natural monopolies. The classic case in the United States is Standard Oil, which controlled about 90 percent of the petroleum industry at the turn of the twentieth century. A number of law suits were brought against Standard Oil. The climax of legislative efforts against Standard Oil was the suit brought by the federal government in 1911 that broke up the company into 33 smaller companies. The government's case indicted Standard Oil under the provisions set forth in the Sherman Antitrust Act. Scholars have identified a number of other historical examples of monopolistic activity.[10]

Critics may point out that there are relatively few historical instances of successful monopolies and cartels. However, when properly appreciated, this point actually bolsters the case against the iron law. As an analogy, suppose it is claimed that killing the CEO of a rival company is a good means to increase profits. As we turn to the historical record, we find that there are very few cases of contract killing of rival CEOs. Rarely, for example, do we find a company like Microsoft contracting the killing of the CEO of a rival company like Apple. However, it is far too hasty to conclude that this is a bad business strategy. On the contrary, when there are heavy punitive sanctions against such activity, and the activity still occurs, it argues that it may well be an effective, albeit illegal, strategy. Similarly, when even a few firms risk legal sanctions in order to engage in monopolistic activity, this argues that such behaviors may well be an excellent strategy for maximizing profits. There are always at least some firms trying to surreptitiously form monopolies even at the risk of heavy fines and punishment. This is a good indication that forming monopolies is a good way to create above-average profits.

Game Theory

The historical record in this respect is polluted: there has been no test of the iron law because anticompetitive activity has long been illegal in industrially advanced nations. Ideally, to assess the effect of anticompetitive legislation, we would run "alternate histories" where capitalism develops in the absence of such restrictions. But just as politicians are not going to lift laws on homicides of CEOs to satisfy our scientific curiosity, neither will they lift laws designed to thwart monopolistic activity to satisfy our scientific curiosity.

Economists work around the limited historical data by using game theory to analyze economic behavior. One thing that economists have studied with game theory is whether predatory pricing, one of the most commonly used tools to support monopolistic activity, is economically rational. In a review of the economic literature, Patrick Bolton, Joseph F. Brodley, and Michael H. Riordan suggest that economic game theory and history confirm that predatory pricing is a rational business strategy:

> It is now the consensus view in modern economics that predatory pricing can be a successful and fully rational business strategy; and we know of no major economic article in the last 30 years that has claimed otherwise. In addition, several sophisticated empirical case studies have confirmed the use of predatory pricing strategies.[11]

The reason predatory pricing can be rational is explained by Paul Milrom:

> Thus, for example, a firm in an industry with rapid product change might cut prices sharply in answer to new entry in order to discourage the new entrant from continuing an active product development program. Whether the entrant attributes its lack of profitability to its high costs, to weak market demand, to over-capacity in the industry, or to aggressive behavior by its competitor, it will properly reduce its estimate of its future profits. If its capital has other good uses, this might lead it to withdraw from the industry. If not, it may nevertheless be dissuaded from making new investments in and developing new products for the industry. At the same time, other firms may be deterred from entering the industry. If *any* of these things happen, the predator benefits.[12]

The upshot of using the anticompetitive strategy of predatory pricing is that it is fully rational, economically. The importance for our purposes is this: if the iron law is true, it should not be a fully rational economic strategy. For if the iron law is true, it is a law of nature and cannot be broken. It would be irrational to sacrifice profits in the short term, as predatory pricing strategy requires, as a means to achieve an impossible goal. The only outcome, if the iron law is true, must be a reduction of profits and hence a nonrational business strategy. Since economic theorists agree that predatory pricing is a fully rational economic strategy (at least where it is not illegal), the iron law cannot be true. If we are to avoid the monopolistic and anticompetitive effects of predatory pricing, then antipredatory pricing must be maintained and managed by legislation: it is not the inevitable effect of an iron law of nature.

More History: Positive and Negative Legislation against Cartels

As noted, cartels are another way for multiple firms to yield monopoly prices. Cartel monopolies often have an advantage over single-company monopolies: with several companies in a market, there is at least the appearance of competition. Indeed, there are a number of historical examples of monopolies setting up a number of companies to appear to be competitors for this very purpose. The true ownership of the company is hidden in a labyrinth of paperwork. Standard Oil used this trick: in some markets, it appeared that there were several competitor companies up against Standard Oil, when, in fact, all the companies were ultimately owned by Standard Oil. Consumers were treated to the illusion of competition.

Those who believe in the iron law sometimes point to the relative infrequency of effective cartels. Cartels often operate by using self-imposed quotas, yet there are huge incentives for businesses to cheat on any agreement that cuts production. Obviously, if the higher than competitive market rate for a product is maintained by firms agreeing to keep production at a certain level, then any one firm can increase its profits by surreptitiously selling more than the agreed upon production quotas. The difficulty in monitoring and enforcing cartel arrangements is cited as evidence of the iron law exerting its effect.

What critics have failed to notice is how this is not any different from other agreements that companies have with each other. There is a heavy incentive to cheat in any business deal. Company A might do well to order 1,000,000 widgets from Company B and agree to pay 90 days after delivery. Company A could potentially make more profit by refusing to pay company B by getting the next shipment of widgets from company C—a competitor of B. Of course, to this, it will be replied that Company B has recourse with the law: it can take legal action against company A in an attempt to recover the agreed upon price.

Why will not the same strategy work in the case of agreements to form a cartel? The answer, in part, is that cartels are illegal in most industrially advanced countries. Importantly, there are in fact two kinds of regulatory restrictions. There is what I want to call "positive" legislation against the forming of cartels, for example, laws against planning cartels, drawing up of contracts for cartels, and so on. And there is "negative" legislation against the forming of cartels, where courts refuse to enforce cartel agreements between companies. In jurisdictions with negative legislation only, companies can conspire to form cartels. They can write up agreements, set prices, and set production quotas. The courts, however, will not enforce the contract if one or more parties break it. In contrast, in jurisdictions with positive legislation, it is illegal to conspire to make such contracts in the first place. Positive legislation may be necessary to stop some forms of cartel behavior, but negative legislation has a similar effect: in refusing to uphold cartel contracts, the courts are trying to influence the behavior of economic actors by making it much, much harder for cartels to flourish.

What should we make of such legislation against cartels? Positive and negative legislations against cartels certainly make sense from a consequentialist perspective. Cartels drive up prices and reduce efficiency in the market, so there is a clear public interest in discouraging their formation by not enforcing cartel agreements and, indeed, by introducing positive legislation against any such agreements.

However, neither form of legislation is consistent within a Lockean framework. Imagine several companies consider forming a cartel in a jurisdiction governed by Lockean justice. In order to sidestep the problems we noted with the Lockean proviso, let us grant that the holdings of the companies proposing to form a cartel are justly acquired. (As questionable as this assumption is, making it favor the Lockean, there is hardly room for complaint from our opponent.) The question then is to resolve whether their agreement respects justice in transfer, and it is hard to see why not. All parties to the cartel agreement freely sign on, anticipating higher profits. The cartel agreement proposes quotas in production and minimum prices. The agreement also specifies what to do in the case of potential competition, for example, use predatory pricing. Any attempt to positively legislate against a cartel agreement violates the company's Lockean rights to manage their property as they see fit. Property owners are supposed to be free to contract with one another. Where the state limits the liberty of property owners to contract with one another in the public interest, the state is using exactly the sort of reasoning that is forbidden in the Lockean realm. Again, as Nozick understands it, "there is no moral outweighing of one of our lives by others so as to lead to a greater overall *social* good. There is no justified sacrifice of some of us for others." A Lockean must see positive legislation against cartel contracts as just this: weighing of the social good against the interests of the contracting cartel partners. Similarly, if the state picks and chooses which contracts to enforce, refusing to enforce those that go against the public interest is again the same sort of reasoning. In refusing to uphold a cartel contract, as would be required by negative legislation, we would be sacrificing the interests of the injured party or parties in a cartel contract for the public good. So, if businesses X, Y, and Z contract to reduce their output by 10 percent to drive up the price of their widgets, and company Z cheats, X and Y should be able to sue Z in court. If the court refuses to hear the case because such cartel agreements do not promote the social good, companies X and Y can say that their interests are being sacrificed for the social good. This sort of sacrifice is supposed to be prohibited in the Lockean framework.

There is a long history of violating Lockean rights when it comes to cartel activity. Positive legislation against cartels in the United States is relatively recent: the Sherman Antitrust Act was born in 1890. The English common law tradition of negative legislation predates the Sherman Antitrust Act by hundreds of years: English courts have long held "restraint of trade" agreements null and void where such agreements are against the public interests unless they are "reasonable" to protect the purchaser.

What this means is that history is bunk when it comes to the question of whether monopolies will naturally develop in a capitalist economy. The historical record of capitalism is spoiled by governments refusing to enforce contracts between parties attempting to create monopolies. Consider Milton Friedman's thoughts on monopolies, "A monopoly can seldom be established within a country without overt and covert government assistance in the form of a tariff or some other device. It is close to impossible to do so on a world scale."[13] Friedman makes this claim absent of any historical evidence on his side. For even if we grant that monopolies have seldom formed in a country, this is not the right sort of evidence to support his claim. The evidence needed, and not provided, is whether monopolies would form if governments enforced contracts between parties attempting to establish a monopoly. The same point applies to the idea that it is close to impossible to form monopolies on a world scale. Here again, the proper test would be to let international courts enforce agreements between producers to form monopolies.

A World with Unrestricted Private Property Rights

Since there is no unpolluted historical record to adjudicate the question of whether natural monopolies will form, we will have to try and determine what would happen in different situations. Suppose there were no government restrictions on private property rights in the name of competitiveness. Imagine the US government, in both its legislative and judicial branches, promises not to interfere in the market if firms use predatory pricing to eliminate competition, or form cartels, and imposes no restrictions on voluntary acquisition resulting in natural monopolies. Thus, in our little thought experiment, we are imagining there is no negative legislation, but governments still perform their vital Lockean function of enforcing contracts. Businesses that signed cartel agreements would expect the agreements to be upheld by courts of law.

Of course, we do not know exactly what would happen, but it is easy to discern the general trajectory the economy would take. In the first instance, there would be a mass rush of mergers and the formation of cartels within different markets. Walmart, for instance, might buy out Target and Costco, two of its biggest competitors. (Target competes in the discount department store category and Costco competes with Walmart's Sam's Club division in the warehouse club category.) AT&T and Verizon control about 70 percent of the wireless market at present in the United States, so we should expect a merger between them and also some of their smaller competitors. The number of wireless providers would shrink. We should expect massive consolidation in other industries as well.

The reason we should expect buyouts is that the logic for investors is inescapable. For example, suppose you owned stock in Lowe's, the second-largest home improvement retailer in the United States, with 16.7 percent of the market. The largest retailer, Home Depot, has 19 percent of the home improvement market. At a shareholders meeting, there is a proposal to form a new company, Home Lowe, which would merge the two companies. Should you, as an investor, vote for the merger? The proposal is to trade stock from the old companies for stock in the new company. The current value of Home Depot stock, let us suppose, is $10, and $10 for Lowe's. Stock in the new company will be purchased for $20, so two stocks of each of the old company will buy one stock in Home Lowe. The question for you as an investor is whether the new company is likely to be more profitable in the future than Lowe's if Lowe's were to not merge. It would be an easy decision to vote for the new company. In part, the higher expected profits would come from greater efficiency: where two distribution systems are used now, one could be used. Also, some stores that are underperforming could be closed. For example, in markets where there is a Lowe's and a Home Depot in close proximity, one could be closed. With nearly 36 percent of the market, Home Lowe could also raise its prices with less fear of losing market share. Also, as the proverbial 800 lb. gorilla in the industry, Home Lowe could throw its weight around quite effectively. Home Lowe would be in a much better position to effectively use predatory pricing. Imagine Home Lowe has 90 percent of the market share in region A and only 25 percent in region B, 60 miles away. The competitive price of 2x4 lumber, we will assume, is $2.00. In region A, it is $2.30 and in region B, Home Lowe charges $1.70. (The price difference is not so great that it would be economical to make the 120 mile round trip to save 60 cents on a 2x4.) Home Lowe is able to average a competitive return on selling 2x4s, even though it loses money on them in region B. The local competitor in region B is a family-owned lumberyard that has been in business for 100 years. Its choices are not appealing: either the family must lose money on each sale of 2x4s, or give up selling a staple of the industry. In either case, its profitability will be drastically affected. Either it will lose money or lose market share and will eventually die or be forced to sell to Home Lowe. The Home Lowe gorilla could use this same strategy to crush many of its competitors. Menards, for example, is often said to be the third-largest home improvement retailer (but it is not publically traded, so it does not have to report its sales or earnings). Home Lowe would be approximately 15 times the size of Menards. What would you say if you were the owner of Menards when Home Lowe came inquiring about buying your business? Assuming your reasoning was based on maximizing your monetary interests, you would have to be crazy to try to go toe-to-toe with a competitor 15 times your size. Indeed, Home Lowe could probably drive down the asking price for Menards with the threat of a predatory pricing strategy that could drive the smaller competitor out of business. As the owner, you should reason that it would be better to take less than the business is worth in a competitive market than risk losing everything in an all-out war with a much larger competitor.

Stepping back for a moment, we can see why consolidation would be a natural result. There are two general strategies for gaining market share: businesses can outhustle or outmuscle their competitors. Textbook apologies for capitalism emphasize the outhustling means: you too can be a rags-to-riches story if your company offers better service, better prices due to efficiency, and/or better selection than the competitors. Outmuscling would mean simply having more capital than a competitor to either buy out the competitor or drive the competitor out of business by predatory pricing. The reason the latter is not emphasized is because the most flagrant forms of outmuscling are outlawed. As we have noted, predatory pricing and the formation of cartels is outlawed in all industrially advanced nations. In a Lockean world, where both strategies are permitted, businesses that do not use both strategies would be at a competitive disadvantage. The outmuscling strategy necessarily involves consolidation in an industry where larger companies gobble up smaller ones.

After an initial round of consolidation, we should expect even further consolidation as companies look to eliminate residual competition and seek more clout in the market. For example, Walmart, after it buys out Target and Costco, might look to buy out Home Lowe. At present, Walmart, Home Depot, and Lowe's compete on certain items such as major appliances, and gardening accessories, to name but a few. Walmart might find it to its advantage to boost the profitability of its appliance division by buying out Home Lowe.

It is an empirical question of how far this process of consolidation and elimination of competition would go, but there is nothing in principle that would prohibit everything being owned by a single company in a Lockean world. Certainly there is no reason to suppose that competitive markets will be the result of some iron law, and much to support the opposite contention.

Back to the Lockean Argument Against Taxes

Recall the Lockean argument against redistribution of wealth. It has two fundamental premises:

Premise 1: We ought not to violate a person's rights.

Premise 2: If we redistribute wealth (as in the BIG proposal), then we violate a person's rights.

Conclusion: We ought not to redistribute wealth.

Our argument demonstrates a dilemma for a Lockean defense of capitalism. On one hand, if we are serious about never violating Lockean rights, then an essential condition of capitalism will be threatened, namely, competitive markets. On the other, if the premises are weakened enough to support capitalism, then the argument does not provide a principled defense against redistribution.

For example, one way to weaken the argument would be to deny Premise 1. We might say we ought to violate a person's rights only when there is a pressing public need to do so. We could then add that violating Lockean rights to stop the formation of cartels and monopolies does serve a pressing public need, namely, the maintenance of competitive markets, and hence, a necessary condition for capitalism. But this would allow at least the possibility of defending redistribution along the same lines: taxes serve a pressing public need. I am not saying that the argument has been made; only that weakening Premise 1 in this fashion allows at least the conceptual possibility of such a defense. We will attempt such a defense in later chapters. At this point, we simply need to notice that weakening the argument in this manner allows the possibility of such a defense.

Another way to weaken the argument would be to say that property rights are not unlimited. This understanding would allow, for example, a community to expropriate property to develop a new highway. Here the community would say to the recalcitrant homeowner: we have not violated your right to private property because your right is limited by pressing public need. Property rights are not vacuous on this conception because they would still limit others from acting against your property interests when a pressing public good was not at stake. It would stop individual A from expropriating individual B's home simply because A wants the property with a nice view. This would also allow the community to stop monopolistic activity: the community could say to would-be monopolists that their rights do not include forming monopolies.

The problem is that this weakened understanding of property rights makes Premise 2 problematic: it is no longer clear that the case of redistributing income in the form of taxes necessarily violates a person's property rights. It will depend on whether there is a pubic need for such redistribution.

Conclusion

The hallmarks of capitalism are private ownership of property and competitive markets. As we have seen, an unlimited right to private property is in tension with the preservation of competitive markets. This leaves the rights-based defense of capitalism with a dilemma.

On one hand, the Lockean might say, "so much the worse for capitalism then if it is inconsistent with Lockean rights." We might think of this position as "Lockean fundamentalism" because the position is willing to support Lockean rights no matter the consequences. I take it that this position has so little appeal it does not need much comment. A society where capital is controlled by so few that competitive markets are by-and-large wiped out does not promise freedom. Rather, it would mean something akin to economic slavery for most, for the few with control over capital would be able to dictate economic terms to the rest. Such a situation would be better for the few who hold the reigns of economic power, but much worse for everyone else. Naturally, the Lockean fundamentalist will object to any sacrificing of the interests of these few individuals for the social good.

More plausibly, apologists for capitalism will concede the need to limit individual property rights for the sake of the public good of competitive markets. While this position is much more plausible as a theory of economic justice, it at least opens the door to further redistribution. I say "further redistribution" since limiting Lockean property rights in the name of competitive markets redistributes wealth from the counterfactual alternative of Lockean fundamentalism. This redistribution is done in the name of the public good.

So capitalists must share a fundamental premise with this work: property rights must be limited in the name of the public good. The residual disagreement is how the public good is best realized and maintained. Apologists for laissez-faire capitalism may argue that the public good is best realized by keeping government interference into competitive markets to a minimum. As we have just seen, there will be some restrictions on the sorts of contracts permitted by economic actors in order to maintain competitive markets, but governments should not go beyond this minimal function. The argument of this work is that this is not enough: the public good is better promoted by offering BIG. In a sense, then, this is an in-house disagreement because the fundamental premise of adjudicating economic justice (at least in part) in terms of the public good is common ground.

We may sum up our response to "taxes are theft" thus: without further support, the claim has no bite, for it merely begs the question. It assumes a certain view of economic justice rather than argues for it. When the taxes are theft slogan is backed up by the view that we have an unlimited right to private property, the argument has too much bite: it devolves into Lockean fundamentalism. The slogan is, no doubt, for scoring political points for people like Boehner, but it does not survive philosophical reflection.

CHAPTER 5

Peace, Robots, and Technological Unemployment

Introduction

In the previous chapter, we saw that capitalism itself requires redistribution to maintain capitalism. In this chapter, we will see that redistribution is also required for an important social good: peace.

Peace has long been a social goal accepted by political philosophers of all stripes.[1] Today, we tend to think of peace in international terms: peace between countries. Yet, historically, most political philosophers have concentrated on peace within a society: nonviolence between individuals or among factions within a country. Plato and Aristotle knew civil war in ancient Greece and wrote treatises outlining theories of government that would reduce the risk of violent upheaval. Hobbes and Locke wrote during the period of the English civil wars, which, no doubt, explains the great emphasis they placed on peace. Of course, the Arab spring and the recent "Occupy Wall Street" protests remind us that the possibility of civil unrest and civil war is ever present.

Yet, at least since Karl Marx, the issue of distributive justice has eclipsed peace as being of paramount importance in political philosophy. In the nineteenth and early twentieth centuries, utilitarianism inspired many social and economic reforms in Great Britain. In analytic philosophy, distributive justice has been at the fore of political philosophical debate since the publication of John Rawls's *A Theory of Justice*.

However, the two subjects of distributive justice and peace are never too far apart. Plato realized that the question of who owns what could be the source of social unrest. He proposed that the idyllic state would be ruled by propertyless philosopher kings—only the governed would own wealth. Plato thought that such a division would do much to keep peace and good order in society. The issue of distributive justice was also not absent in Hobbes' work, and it is discussed to a much greater extent in Locke's work. Although Marx placed greater emphasis on distributive justice than any of his predecessors, the issue of peace is not absent in Marx's work either: he famously predicted a violent end to capitalism. With communism, a new era of peace was to ensue across the globe.

This chapter argues that the concerns of the ancients and the moderns are our own. Peace and distributive justice are intimately intertwined. We are at a point in history where we have very good reason to suppose that there will be a massive shift in the economy due to computers and robotics. If collective action is not taken, we should predict this paradoxical result: the cheaper the products, the fewer who can afford them. This is supported by the thought that with computers and robots replacing human labor, we should predict a drop in the prices for goods and services. When a shipping clerk making $15 an hour is replaced by a robot that costs 30 cents an hour to run, we should predict a reduction in prices charged to consumers. However, as workers are replaced by computers and robots, they will be deprived of a source of income. With fewer people participating in the paid economy, fewer people will be able to afford the goods of a massively productive economy.

The issue of peace, then, comes to the fore when we ask how people will deal with this massive shift in the economy. Suppose, for example, there is a net change such that 15 to 50 percent of the workers are made redundant by advances in computers and robotics in the next 25 years. How do you think people will react? Are they going to sit idly? With gaunt smiles from starvation, will they simply congratulate others on their ability to compete in the new economy? As the economy shifts dramatically, we have good reason to fear that the ensuing protests will not be as small or as civilized as the Occupy Wall Street protest if nothing is done.

The remedy is BIG: BIG will help ease us peacefully into the new economic reality. In other words, the argument is that BIG is an effective means to thwart the threat to peace precipitated by the robotic revolution.

Technological Unemployment

One way to analyze the threat to employment from robotics is to consider all the jobs presently done by humans that could soon be performed by robots. Let us fast-forward 20 years and imagine you need to do a little shopping. You go to your local Walmart and pick up a few groceries and other household items. On the way to your office, you realize you forgot to buy an electric razor, so you order one through Amazon on your cell phone. Your trip to the office is to pick up a book you are supposed to review. When you get home, you are not surprised to see that the razor from Amazon actually beat you to your doorstep. As you sit down to dinner with your clean-shaven face and begin to read your new book, you marvel that, in all likelihood, all the products you are using were untouched by human hands other than your own.

How is this possible? The answer, of course, is robotics. The book you are to review was run off and bound by the licensed printer at your university. Yes, you could have received an electronic version right in your tablet, but you are old-fashioned and still like the feel of a paper book in your hands. Hence you had to use one of the few printing and binding machines that you know of, the one at your university. The razor was made at an entirely robotic factory and shipped robotically to an Amazon distribution center. When your order was placed, it was robotically packaged and sent out in a small robotically driven helicopter that dropped the package right at your doorstep.

Your trip to Walmart also did not require human touch (other than yours). Gone are the days of waiting in line to have your items scanned and bagged. As soon as you put an item in your shopping cart, the embedded RFID (radio-frequency identification) tag was read and the item automatically charged to your account. Gone too are the small army of human shelf stockers. This job is now done robotically. Robots are also in use at every step in the distribution and production sequence. Robots packed and drove the food to your local Walmart. Robots also were used to grow the food on the farm, to pack them, and to ship them.

I expect two quite different reactions to this little description of consumption in the year 2035: some will think this is wildly implausible because it attributes too much to robotic developments; others will find it wildly implausible because it attributes too little to robotic development. The latter reaction is probably closer to the mark, but as the former is inconsistent with what is argued here, we shall focus on it.

What may be unnerving for those who do not follow robotic development closely will be not how much extrapolation from our current technology is required for robotics but how little. Some of this is already a reality. Part of our little story involved print-on-demand technology. But print-on-demand robots are already functioning, and by 2035, this will be considered ancient technology.

The idea that the electric razor might reach your hand untouched by any other human is only a small extrapolation from current technology. Recently, Philips Electronics opened a factory in the Dutch countryside that uses 128 robots and one-tenth the human labor as a counterpart of their factory in China.[2] The robots work with greater acuity and dexterity than is possible for an unaided human; for example, one robot bends a connector wire in three places and, guided by video camera, slips the bent wire into holes too small for the human eye to see.[3] These robots are able to do such incredible feats at such rapid rates that they must be enclosed in glass cages, as their rapid speed can endanger the few humans working in the factory. Of course, robots are capable of working 365 days a year, 24 hours a day. The new factory has made obsolete the hundreds of Chinese workers who assemble razors in China the old way, using human labor.

Not only is manufacturing being revolutionized by robotic workers, but also is the shipping and receiving industry. We are on the cusp of being able to get the inventory from factory to warehouse robotically. In some advanced warehouses, Kiva robots are directed by a computer program to select inventory from the warehouse shelves and bring it to human workers who actually place the items into boxes for shipping. On the transportation end, Google has software for driverless cars. Of course, legal restrictions still require a human driver to be present who is capable of taking over the wheel, but the software is already so good that experimental vehicles have gone hundreds of thousands of miles without any human intervention. The safety record of this software already exceeds that of the average human driver. It is not hard to imagine that human-driven vehicles may be illegal in 20 years, simply because they would be more dangerous in comparison to robotically driven vehicles. Thus, it only requires the barest extrapolation on today's technology to imagine robots packing orders at the factory, driverless trucks shuttling inventory between factory and warehouse, and robots packing and delivering orders straight to the doors of customers. Amazon recently demonstrated a helicopter delivery drone that could potentially make 30-minute deliveries possible, with the prediction by its CEO, Jeff Bezos, that the technology could be deployed within five years.[4]

An obvious analogy here is the great reduction in the workforce as a percentage of the population devoted to agriculture. In 1790, there were about 3.5 million farmers, which was 90 percent of the population. Farm workers now make up 1.6 percent of the population, comprising 5 million workers.[5] Poignantly, even this small remaining farm workforce will be replaced in large measure by "farmbots." Small prototypes of farmbots being tested in the field do such things as plant seeds, pull weeds, and harvest in swarms. Although they are not ready for prime-time deployment at present, it is already possible for robots to do planting, maintaining, and harvesting jobs in the laboratory. In all likelihood, in 20 years, robots will greatly outnumber humans on farms in the United States. Agriculture is yet to see the last reduction in its workforce.[6]

We will see a radical reduction in the need for human employment not only in manufacturing, distribution, transportation, and agriculture, but also in more "cerebral" professions. There are medical programs that outperform even experienced physicians in diagnosing disease.[7] Surgical robots are also being developed. In both cases, the physician and the surgeon are not completely replaced, but it is easy to anticipate that the need for both on a per-capita basis might drop dramatically. For example, up to 60 percent of visits to family physicians are for upper respiratory infections, which can be easily diagnosed by a computer program.[8] The claim here is not that these professions are going to be completely replaced by robotics in the next 20 years, but, rather, that robotics will increase the efficiency of physicians and surgeons so that fewer will be necessary per capita. The title of a recent CNN article summarizes the trend, "Technology will replace 80 percent of what doctors do."[9]

It seems that hardly a week goes by without some headline about robotics taking over jobs in a new field. On the day I had almost finished a draft of this chapter, a *Wall Street Journal* headline appeared with this intriguing title, "Robots vs. Anesthesiologists: J&J New Sedation Machine Promises Cheaper Colonoscopies; Doctors Fight Back."[10] About a billion dollars a year is spent on sedating patients for colonoscopies. The robotic anesthesiologist developed by Johnson & Johnson promises to take over much of the labor of anesthesiologists. Physicians actually performing the colonoscopy typically charge in the $200–400 range for each procedure. Anesthesiologists charge an additional $600–2,000. The robotic anesthesiologist, now approved by the FDA, would work for a fraction of the cost, at $150 or so per procedure.

Medicine is not the only high profile profession under siege of robotics, there are computer programs operating today that can perform legal research faster and more effectively than well-trained lawyers.[11] We need not imagine a future where robotic lawyers stand up in court to give an eloquent defense of the accused to comprehend that the need for human labor in the legal profession will decrease.

The so-called oldest profession should also worry about the reduced need for human labor. Sexbots are available now with several different "personalities," capable of performing a number of different sexual acts. The reason they have not penetrated the market further, as it were, is due to their price, at least partly. They typically retail in the neighborhood of $10,000. As the price drops, there is every reason to suppose sexbots and other robots will further replace more human labor.[12]

Part of why the robot revolution is still invisible to many, despite much recent press, is that we are still suffering a hangover from earlier expectations. There was a lot of optimism in the 1960s that general purpose humanoid robots might be a reality in the 1980s. This promise, obviously, was not fulfilled. Instead, what we have are far more limited special-purpose robots. Presently, robots are designed for very specific tasks; for example, most of the robots sold for floor cleaning are specialized for vacuuming the floor or washing the floor. The idea that we would have a general purpose house servant like "Rosie" from the Jetsons cartoon is a long way off. Still, once our expectations have been retrained, we can see why the robotic revolution is inevitable: robots are getting incrementally cheaper and better every year.

Robotic vacuum cleaners like Roomba have vastly improved in the last decade. When they first reached the mass market, they were designed to clean a single room, were prone to fall down the stairs, and used to get stuck in corners. They also needed to be recharged by humans. Now, many such robots can vacuum a whole house unaided. When the batteries run low, the robots return to their "feeding stations" where they recharge. They are by no means perfect but certainly much better at vacuuming than your average teenager. They also work cheaper and do not complain.

Another example of robotic progress is Baxter from Rethink Robotics. Baxter is an industrial robot designed by Rodney Brooks, inventor of the Roomba robot. Let us consider the cost first. Unimate is usually credited with the installation of the first industrial robot in 1961.[13] This robotic arm worked at a General Motors factory with hot die cast metal sorting and stacking. Unimate sold the robot at a loss: it cost $65 million to make and Unimate sold it for a paltry $18 million. Baxter costs more than 1,000 times less, retailing at $22,000. Even compared to many of its contemporary competitors, Baxter is a giant leap forward. Often the price of a robot is a fraction of the total cost of its operation. For example, a typical industrial robot that costs $100,000 at present might use an additional $400,000 in labor fees to have programmers write and debug code to instruct the robot how to perform its task. Baxter, in contrast, can be trained by factory workers: it is simply a matter of guiding Baxter's arm to show it what needs to be done. Baxter learns by doing, rather than having new code input. This makes the lifetime cost of Baxter cheaper than many of its competitors ($22,000 versus $500,000). Robots like Baxter will revolutionize industrial production.

As mentioned above, the robotic revolution is being spearheaded by specific purpose-built machines. And because of this, there are very few who are able to see the revolution in all its clarity, as it requires knowledge of developments in a number of specialized domains. For example, when a reporter asked Rodney Brooks, inventor of both the aforementioned Roomba and Baxter robots, how long it might be before robots could replace McDonald's workers, his response was very telling. Brooks claimed that "it might be 30 years before robots will cook for us."[14] His reasoning for this prediction is also interesting, "In a fast food place you're not doing the same task very long. You're always changing things on the fly, so you need special solutions. We are not trying to sell a specific solution."[15] I cannot help but wonder whether Brooks has ever worked at a fast-food restaurant. As a former McDonald's employee, I can attest to the repetitive nature of the work.[16] I would describe working there in exactly the opposite way: you do the same task for a very long time with little variation in the routine at each station. Interestingly, there is already a robotic hamburger maker available from Momentum Machines.[17] It will cook up to 360 hamburgers an hour, plus cut fresh tomatoes, lettuce, and pickles. Or consider Kura, a sushi restaurant chain in Japan that uses robotics to lower its labor costs.[18] The fact that a world-class roboticist like Rodney Brooks has underestimated the robotic revolution is revealing. The reality is that we are almost at a tipping point where robots are cheaper even in an industry known for its low-cost labor.[19]

Indeed, what is disturbing is that even in China, with its notoriously low wages and harsh working conditions, there is a move to robotics. The chairman of Hon Hai (also known as Foxcon), manufacturers of Apple's iPhone and other electronic devices, announced a few years back that the company's goal is to have a million industrial robots in use by 2014.[20] The plan has hit some snags but is still proceeding at an aggressive pace.[21] Some analysts say the price of robots is still too high for it to make economic sense for Hon Hai, suggesting that the price per robot would have to fall to $25,000 from their current $50,000–$200,000 level. Interestingly, this is the price of the aforementioned Baxter. So, again we are reaching a tipping point where even in a low-wage country like China, it makes economic sense to replace human workers with robots. Indeed, as we noted, Philips Electronics has already found it more economical to set up a robotic factory in Europe than have electric razors made in China with cheap labor and then shipped to Europe.

The logic behind this move is explained by Hon Hai Chairman Terry Gou: "Hon Hai has a workforce of over one million worldwide and as human beings are also animals, to manage one million animals gives me a headache."[22] Terry Gou added that he wanted to learn from Chin Shih-chien, director of the Taipei Zoo, regarding how animals should be managed. As offensive sounding as this is, it is hard to deny the logic of his thinking from a ruthless business perspective. Humans are expensive machines to run and maintain. The working conditions of Hon Hai regularly make the headlines because, as noted, they are the major manufacturer of Apple products. It is alleged that the working conditions at Hon Hai are terrible and that there have been a number of suicides as a result. Whether this is true or not, it is clear that negative publicity is not something Hon Hai desires. Even if humans could compete with robots in terms of work produced per hour, the extra "headaches" of managing "animals" is surely going to tip the scales in favor of robots. Focusing just on the economics of the issue, if the price of robotic labor and human labor is even, then robots will be the preferred choice, leaving humans in the dust.

The Economists versus Chicken Little

It cannot be stressed enough that the argument is *not* that machines will completely replace all human labor. This may come to pass one day. However, there are good moral reasons why it would be wrong for robots to replace humans in all occupations.[23] Long before this transpires, robots will partially replace humans in the workforce. So the argument requires only the relatively modest claim that robots will *reduce* the need for human labor below full employment.

It may help to put some numbers to these claims. The height of unemployment consequent to the great recession of 2008–2009 was about 9.9 percent in 2010.[24] If technological unemployment continues to occur, in the next ten years the rate of unemployment will consistently remain higher than this and then it will continue to grow.

In other words, the reason for the higher unemployment rate will be due not to the cyclical nature of markets or subprime mortgages, but because of huge increases in automation.

To this conjecture, it is often objected that economies have also found jobs for displaced workers in the past; for example, when backhoes automated the work previously performed by hundreds of workers on a single worksite with shovels, many laboring jobs were lost. But economic growth spurred new employment opportunities for the displaced laborers, including work in factories making backhoes, backhoe mechanics, and backhoe salespersons.

Pressing the objection, it might be suggested that the analysis here is mere Luddism. The Luddites of the early nineteenth century reacted against the mechanized machinery of the industrial revolution. Skilled artisans in the textile industry were replaced by machinery and less skilled labor was required to operate the machinery. In response, the Luddites wrecked machinery, killed capitalists, and battled with the British army.

In hindsight, it is easy to sympathize with their plight if not their prescription. These artisans had much to fear. It is true that the economy as a whole benefited from the reduction of the price of textiles, but most of these workers did not reap a commensurate reward. Their jobs were permanently lost to automation and their particular skill set did not position them well to compete in a new economy. Imagine their plight when the workers, who put in the requisite years to learn their craft, found that their skill was no longer needed and that they would have no way to look after their families. At the micro level of the individual worker, this is very sad. At the macro level, however, we can see the benefit to the entire economy.

The same thing that happened to these workers in the textile industry happened in many other professions. The objection to these developments, then, can be summarized by saying that the history of the modern world is one of workers made redundant by machines. Whenever workers are made redundant, disciples of Chicken Little have suggested that the employment sky is falling. Each time, however, the prediction of massive unemployment as a permanent feature of the economy has proven false. Yes, there is pain as the economy readjusts, but eventually workers find new jobs, often in entirely new industries. This line of rebuttal to technological unemployment is nicely summarized by the economist Alex Tabarrok:

> I am growing increasingly annoyed with people who argue that the dark side of productivity growth is unemployment. *The Economist*,

> which ought to know better, says we are overproductive. *CNN Money* discusses the problem of productivity, the President blames productivity growth for unemployment. Even someone as sophisticated as Brad DeLong writes "with productivity surging, it's hard to be pessimistic about GDP growth, but it's easy to be pessimistic about unemployment" which seems to suggest that if only productivity growth were lower, employment would be higher.
>
> And yet the "dark side" of productivity is merely another form of the Luddite fallacy—the idea that new technology destroys jobs. If the Luddite fallacy were true we would all be out of work because productivity has been increasing for two centuries. Sure, some say, that may be true in the long run but what about the short run? Even in the short run there is no necessary connection between productivity growth and job loss. In the computer industry, for example, productivity growth has led to falling prices and a bigger not smaller industry. If demand is inelastic then productivity growth can create short-term unemployment, especially at the level of the industry experiencing the growth—less likely but not impossible is that productivity growth leads to short-term economy-wide unemployment.[25]

I will refer to this objection as "the economists' objection" simply for ease of reference. I do not mean to suggest that all economists endorse this idea, nor do I mean to suggest the idea is proposed only by economists. Indeed, the belief that the economy will always generate enough jobs seems to be well entrenched: every year I hear college freshmen state it with the same unflinching faith, just like economists such as Tabarrok.

Notice that the economists have the more extreme position. They are committed to the idea that the economy will generate full employment. Chicken Little wins if there is anything less than full employment due to technology's advancement. Despite Chicken Little's horrendous track record of failure in predicting technological unemployment, I think the prediction of technological unemployment is sound.

The Nature of the Debate between Chicken Little and the Economists

Thus far, I have merely rehearsed the outlines of a familiar debate. It is easy to get the impression that we are at an argumentative impasse. On one hand, the case for Chicken Little is based on the observation that computers and robotics are making inroads into so many sectors of the economy: agriculture, mining, construction, manufacturing, retail, professional services, teaching, health care, and food services, to name but a few. On the other, it is hard not to concede that the economists have a very powerful case. Their argument in a nutshell is as described:

Premise 1: People wrongly claimed that automation will result in massive technological unemployment during 1811–2014.

Conclusion: People who claim in 2014 that robotics will result in massive technological unemployment are wrong.

The seeming impasse noticed here is exacerbated by the problem that we discuss here. If we press the economists to tell us where the new jobs will be created for workers in the displaced industries to move to, they will (rightly) respond that it is generally impossible to foresee. The worker replaced by the automated threshing machine had no inclination that his grandson would work as an elevator operator. The elevator operator, made redundant by automated elevators, had no idea that his granddaughter would be a cell phone engineer. This makes the economists' position hard to criticize, because they claim that some new, but as yet unknown, sectors of the economy will open up and employ workers made redundant by automation. It is, of course, hard to argue against the unknown.

Despite this, I believe the case can be pressed against the economists. The first thing we should notice is that the economists' argument is an inductive argument based on the general premise that the future will resemble the past. The general pattern of reasoning is straightforward enough. A reason to believe that the temperature will dip below freezing next winter is that it has been dipping every other winter for as long as humans have recorded temperature in these parts. A reason to think the sun will rise tomorrow morning is that it has risen every morning since before humans populated this planet. Similarly, the economists' argument uses the same inductive pattern: every time automation displaced workers in the past, new jobs were found.

A crucial difference between inductive arguments and deductive arguments is that it is possible to accept the premises of an inductive argument but deny the conclusion, whereas with a deductive argument, it is not. Consider this deductive argument:

Premise 1: People wrongly claimed that automation would result in massive technological unemployment during 1811–2014.

Premise 2: If people wrongly claimed that automation would result in massive technological unemployment during 1811–2014, then people who claim in 2014 that robots will result in massive technological unemployment are wrong.

Conclusion: People who claim in 2014 that robotics will result in massive technological unemployment are wrong.

If we accept Premises 1 and 2, then we are logically forced to accept the conclusion. With inductive arguments, it is possible to accept the premises but deny the conclusion. Generally, however, we need special reasons to deny the conclusion, because denial of the conclusion of an otherwise good inductive argument requires denying the idea that the future will resemble the past. Suppose with the assistance of a time machine, you are transported to Pompeii, August 23, 79 CE. Someone tells you that fresh bread will be available in the market the following day, based on the fact that fresh bread has been available in the market every day for over 50 years. This is a very good inductive argument, but still you have reason to deny the conclusion, that bread will be available tomorrow, even while accepting that it has been available for the previous 50 years. Your knowledge that Vesuvius will blow the next day gives you reason to think that the future will not be like the past for the poor city of Pompeii.

I propose to take a similar line with the economists' argument. Although I think there are reasons to deny the premise, we will assume for the sake of the argument that the premise, that automation has not caused unemployment, is correct.[26] I will argue that there are reasons to think that the future will not resemble the past with respect to employment. Accordingly, we have reason to deny the conclusion of the economists' argument.

The Good of Humans in an Economy

To see why the future will not resemble the past, economically speaking, it will help to step back for a moment and ask what role humans play in the economy. On the demand side, things are relatively straightforward: humans are the primary consumers of the economic goods produced in economies. On the supply side, our primary benefit is in the form of labor. What we offer to the economy in terms of labor is helpfully illustrated by comparison with horses:

> There was a type of employee at the beginning of the Industrial Revolution whose job and livelihood largely vanished in the early

> twentieth century. This was the horse. The population of the working horses actually peaked in England long after the Industrial Revolution, in 1901, when 3.25 million were at work. Though they had been replaced by rail for long-distance haulage and by steam engines for driving machinery, they still plowed fields, hauled wagons and carriages short distances, pulled boats on the canals, toiled in the pits, and carried armies into battle. But the arrival of the internal combustion engine in the late nineteenth century rapidly displaced these workers, so that by 1924 there were fewer than two million. There was always a wage at which horses could have remained employed. But that wage was so low it did not pay for their feed.[27]

It is interesting to ask why new career opportunities did not open up for horses after the invention of the combustion engine. After all, if we are to believe that new job opportunities will open up for humans after the robotic revolution, then surely capitalism should have found jobs for horses after the internal combustion engine revolution. The unbridled optimism of the economists seems to suggest full employment for horses too. So why did so many end up at the knackers? And why should not we predict the same thing for human workers?

The answer is perhaps obvious: horses have one main thing to offer the labor market, namely, their physical labor. As the quote from Clark indicates, it is not that physical labor is not valued in the modern economy, it is simply that the internal combustion engine (or electrical engine) can provide the same physical labor much more cheaply.

Horses also have some residual value in terms of what we might call "nostalgia." You can rent a horse and go for a ride. Or you can take a ride on a horse-drawn sleigh or go on a carriage ride through a park. No one pretends that these are the most efficient means to get around. Some people just like being around horses or enjoy fantasizing about a bygone era when horses used to be the primary mode of transportation.

So, horses have two things to offer to the economy: muscle and nostalgia. By and large, their muscle power is not cost-effective in a modern economy.[28] And the demand for horse nostalgia is not enough to keep horses employed at the same rate they were in the early part of the twentieth century.

Humans have three things to offer the economy: brains, muscles, and nostalgia. History shows the inception of two great transformations in the economy. The first, approximately during 1800–1950, was when human muscle power was replaced by machine power.

Thus, starting in the eighteenth century, there was a rapid rise in development of machinery to replace human physical labor. First the steam engine and, later, combustion and electrical engines replaced human and animal labor. Take a simple example. A backhoe can dig and prepare the foundation of a building in a day. A small army of humans with shovels might take a month to do the same thing. In terms of energy output, we can see why this is so. An average human might generate one-tenth of a horsepower over an extended period of time, whereas a typical backhoe might be rated around 100 hp. A backhoe operator has at his disposal the equivalent muscle power of 1,000 average human beings.

Perhaps nowhere is the replacement of human physical labor with machine power more obvious than in agricultural production. As discussed earlier, at the turn of the eighteenth century in the United States, about 90 percent of the population was involved in agricultural production. A mere hundred years later, this fell to 40 percent of the population. One manual-labor-saving device that reduced human labor was the threshing machinery introduced in the late seventeenth and early eighteenth centuries. Threshers separate wheat and other grain kernels from the rest of the plant. Prior to the introduction of the threshing machine, this was done through the use of a flail. A typical flail comprises two sticks of wood joined by a piece of chain. A flail, then, looks much like a nunchaku weapon often seen in Kung Fu movies, although the sticks of a flail tended to be longer. Workers would beat wheat and other grain crops in order to "separate the wheat from the chaff." This was extremely physically demanding and labor intensive. With the introduction of the threshing machine, about 25 percent of the farming work force was made redundant. Again, at the individual level, we should feel sorry for those farm workers who were laid off and unable to find different work due to the invention of the threshing machine. But at the social level, we can see how automated farm machinery ultimately improved the lives of many. No longer must 90 percent of the population work on farms for long hours, low pay, and in dangerous working conditions.

The Second Great Transformation, 1950–2050, will be when computers and robots eventually replace human minds in the economy. Humans can still compete in the area of the mind; but as we have seen, this advantage is dwindling. Robots that work in factories, advanced computers that drive cars in busy traffic or make accurate medical diagnosis, or even do effective law research, are all making inroads into areas where humans once had a unique advantage. There are two reasons for thinking that robots will continue to chip away at our last great advantage in the labor pool. First, the cost of hardware for robotic intelligence is decreasing every year. This is a consequence of "Moore's Law," which states that the number of transistors on integrated circuits doubles every two years. In plain English, computer hardware is getting more powerful and cheaper every year. Second, gains in software development can be distributed virtually for free. A comparison with humans may serve to highlight the difference. Teenagers are no better at vacuuming today than they were ten years ago. It costs about the same to raise them, and each child has to be taught the art of vacuuming. Compare that with robotic vacuums. If the computing power necessary to run the robot's brain cost $100 ten years ago, it now costs just over $3. Programs for robotic cleaners have also improved in the intervening period: robotic vacuum cleaners now are less likely to fall down stairs or get stuck trying to vacuum up a sock. Each time programmers make an improvement, it can be disseminated to current and future robots through a simple download. In short, although we still have one great advantage in the labor pool, our minds, our minds are not getting significantly better or cheaper, whereas robots are improving on both scores.

We can now see why the past is not a particularly good predictor of the future in the case of employment. With the First Great Transformation, humans were squeezed out of jobs that purely used their brawn, but new sectors opened up in jobs that required their brain. In other words, machines now are encroaching into the last area where we have a competitive edge over them. So, unlike the displacement of labor during the First Great Transformation, there is no untapped category of jobs left for surplus human labor to migrate to.

A couple of caveats are in order. First, the distinction between occupations that use one's brawn or brain is best thought of as lying along a continuum, with few (if any) jobs that are the pure forms of either. My job as a philosophy professor is one of the clearer examples of a brain-type job, but even my job requires human muscle. I must, for example, move about and collect student papers, carry books to class, and so on. The brawn involved in my job is fairly minimal, and it is not enough to keep me physically fit by any stretch of imagination. Some of the purer forms of brawn-type jobs include rowing a galley and threshing grain. Neither of these is a pure form of brawn use, as at least some mental activity is necessary for each: following orders about when to row, for example. The claim about the two great transformations, then, may be nuanced by claiming that the First Great Transformation saw a mechanization of jobs that tended to be on the brawn end of the spectrum and now we are seeing the mechanization of jobs that are more toward the brain end of the spectrum.

The second caveat is that humans will still be employed in nostalgic functions. For example, one can take a ride in an elevator in the Young-Quinlan Department Store operated by a human "elevator operator." Of course, at one time, elevators were not automated but run by specially trained elevator operators. The few humans that are still employed as operators appeal to nostalgia; it is certainly not cheaper to run an elevator with humans as opposed to computers. As with horses, we should expect that the demand for human labor based on nostalgia will be pretty weak.

To summarize, the economists' argument says that new areas of the economy will develop to generate full employment. When Chicken Little asks about these job-creating sectors, the economists say that it is hard to predict where the growth will come from. If it were easy to predict where these growth will come from, investment would be straightforward, but it is not. The response to this is that even supposing new sectors of the economy open up, the demand for human labor is likely to be very weak since these new sectors too will be faced with the question of whether to employ robots or humans. The cost advantage will lie with using robots for the most part, and so there will be weak demand for human mental labor in the future, just as the demand for human muscle dropped precipitously in the past.

What If Chicken Little is Wrong?

Despite the strong case for looming technological unemployment, it is worth considering the possibility that Chicken Little is wrong. Chicken Little's dire prediction is based on two claims: (1) that a large number of jobs presently done by humans will be performed by computers and robots in the future, and that (2) new sectors of the economy will not generate sufficient jobs for humans (because machines will supply most of the necessary labor in the new sectors of the economy as well). Interestingly, there is near universal agreement among experts about the first point. The residual disagreement between Chicken Little and the economists is about point two.

It will be worth discussing what the economy will look like if the second point is false and the first is true. For such a scenario, there will have to be massive growth in the economy. Specifically, the growth in economic output will have to be greater than worker redundancy. For example, it might seem reasonable to expect that if 10 percent of the

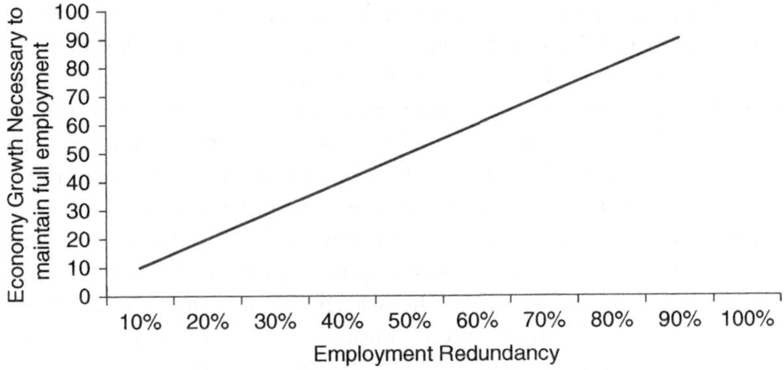

Figure 5.1 One-to-one ratio of economic growth and employment redundancy

workforce is made redundant by robotics, then the economy would have to grow by 10 percent to absorb these workers to maintain full employment. Such a one-to-one relationship is illustrated in the one-to-one graph below (figure 5.1).

However, such a one-to-one relationship is not plausible, as we discuss here. Suppose in an economy with 100 firms, with ten workers per firm, each worker produces 1 widget per year. "Widget" is to be understood as the product of the company, it could be goods or services. This means that the economy produces 1,000 widgets per year. Now, imagine that robotics and advanced computers replace only 10 percent of the work force. It is easy to suppose that the one-to-one relationship must be correct: the economy must then grow by 10 percent in order to absorb these workers and maintain full employment. However, this is not the case. The economy would have to grow by more than 10 percent to absorb these workers. If each firm now only employs nine workers, the number of unemployed will be 100 workers. If the economy grows by 10 percent, this would translate into 10 new firms. In other words, 110 firms each producing ten widgets equals 1,100 widgets per year. But ten new firms would only employ 90 people, because the average number of workers employed has dropped to nine. So, the economy would have to grow by more than 11 percent to get back to full employment. The difference gets more dramatic as the redundancy percentage is increased. For example, if robots replace three out of ten jobs at each firm, then the economy will have to grow by 43 percent to get back to full employment, not 30 percent. For now there will be 300 people who will need to find jobs. Since the firms now only employ an average of seven people, 43 new firms will have to be created to maintain full employment. If robots are able to replace five out of every ten jobs at present, as suggested by Frey and Osborne's detailed study, then 500 people will be unemployed in our economic model.[29] 100 new firms would have to spring up, that is, finding work for 50 percent of the workforce translates into a 100 percent increase in economic output (table 5.1).

The point, in other words, is that new industries themselves will likely use advanced robotics and computers, and so economic output will have to increase faster than the percentage of unemployed to keep the economy at full employment. Figure 5.2 shows the relationship necessary for economic growth and full employment.

Table 5.1 Increase in economic output

Firms	*Employees per firm*	*Total employees*	*Total widgets*	*Percentage of economy in 2014 (%)*
100	10	1,000	1,000	100
111	9	1,000	1,110	111
125	8	1,000	1,250	125
142	7	1,000	1,425	142
166	6	1,000	1,660	166
200	5	1,000	2,000	200
250	4	1,000	2,500	250
333	3	1,000	3,333	333
500	2	1,000	5,000	500
1,000	1	1,000	10,000	1000

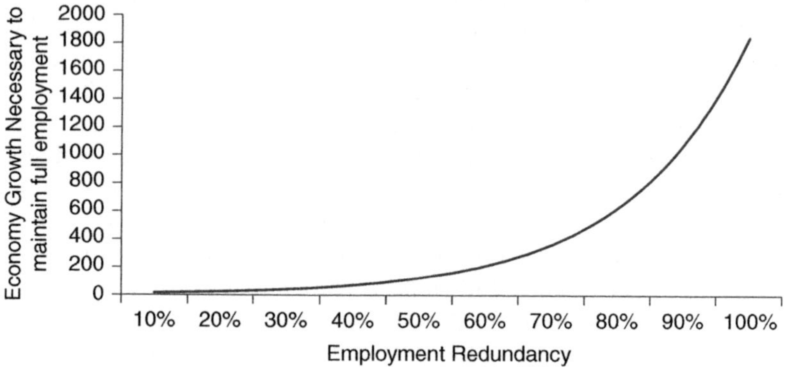

Figure 5.2 Exponential rate of economic growth and employment redundancy

We are now in a position to see why BIG is a smart bet, given the uncertainties in employment creation. The future economy, with massive implementation of robotic workers, may not or may generate full employment for humans. If the former, then there is a clear need for BIG. In such a case, we imagine that workers get simply outcompeted by robots in most areas of the economy. Addressing the needs of such displaced workers is morally and prudentially important. Morally, of course, we ought to care about the plight of our fellow humans. Even for those motivated solely by prudential concerns, BIG would be an efficient way to stop social unrest caused by massive unemployment. Those fortunate to be still employed, in other words, should find BIG an attractive means to avoid the threat of having their heads put on the ends of pikes by angry mobs upset by perceived unfairness of the robotic revolution.

On the other hand, if the economy is able to generate full employment, then the economy will have to grow faster than the redundancy rate to maintain full employment. This means that the economists who predict full employment must also predict a massive growing economy of unprecedented proportions as a logical consequence. That is, optimism about full employment logically requires optimism about a massively expanded economy. In this case, paying for BIG will be comparatively easy as it will be a small percentage of total economic output. As noted in chapter 2, a reasonable BIG for US citizens works out to 11 percent of the economy at present. Using this as our baseline, we can see that if 20 percent of the jobs at present performed by human workers are executed by robots while full employment is maintained, then paying for BIG as a percentage of the economy should fall to 6 percent from 11 percent of the total Gross Domestic Product. If half the jobs are taken over by robotics in an economy with full employment, then the total cost of BIG works out to about 3 percent of this economic future. So, what makes BIG a rational bet is that either it will be urgently needed or easy to pay for (and perhaps both).

Unrest and Unemployment

If there is persistent high unemployment, say over 15 percent, and nothing is done for the displaced workers, then I predict civil unrest. History is only partially supportive of this claim. Certainly, there are historical cases of civil unrest caused by high levels of unemployment and economic desperation. Britain experienced food riots during the 1770s and 1780s. There were general strikes in Britain during 1921 and 1926 due to economic distress, mainly in the working class. The

"Swing Riots" in the 1830s in Great Britain were directly attributable to high unemployment, caused in part by the introduction of mechanized farm machinery, which displaced workers.

The strongest countervailing evidence is the Great Depression. Although unemployment in the United States rose to 25 percent, the history of this period was relatively peaceful. The most notable civil unrest was the use of the US military against veterans during the Bonus Expeditionary Force's protest in Washington. About 17,000 World War I veterans demanded early payment of a bonus promised to them in 1924, which was to be paid in 1945. Their argument was that the Great Depression was sufficient cause to demand that the bonus be paid early by the federal government. The veterans and their families set up camp around Washington. Two protestors were shot by the police, at which point, President Hoover sent in the army to clear out the protestors.

Despite the lack of widespread civil unrest during the Great Depression, there are several major differences to the scenario where robotics replace human labor. One difference is that citizens are less acquiescent today than they were in the 1930s. There is a level of cynicism at present that was not there during the 1930s. People distrust politicians and economic leaders far more today. Modern communication systems, cell phones, Internet sites, text messaging, and so on, make it much easier for the disaffected to communicate with one another and coordinate their activities.

However, perhaps the biggest difference would be people's long-term outlook. Unemployment during the Great Depression was not perceived to be caused, at least in large part, by workers being made redundant due to advances in technology. Scholars typically mention things like the stock market crash of 1929, a large number of bank failures, reduction in consumer demand, foreign trade policy, and so on. During the Great Depression, it was generally believed that the economy would eventually bounce back and jobs would return. People had faith that it was merely the cyclical nature of capitalistic economies. So at least one big difference during the Great Depression was that there was hope things would get better. Suppose you are one of the more than half million truck drivers laid off when driverless trucks take over. If you see doctors and lawyers being laid off too as a result of advanced computers and robotics, what will you hope for in terms of getting your next job? If it is not a cyclical phenomenon but a permanent feature of the economy, then acquiescence will not be fueled by hope.

Alternate Solutions, Increased Oppression, and Welfare State

If economic gains in productivity due to robotics tend to coalesce in the hands of the few, the so-called 1 percent, while more and more workers are made redundant, it is hard to see how civil unrest will not result. Almost certainly before widespread rioting ever occurs, policies would be formulated to deal with the problem. As consequentialists, we should choose the policy alternative that best promotes aggregate utility. So, we should consider how the BIG solution compares with the alternatives.

One possibility would be to simply beef up security. If a large section of the population is agitated because they cannot find jobs while wealth gets concentrated in the hands of fewer and fewer, then an obvious solution is to increase the scope and power of law enforcement services. It would be an easy matter to create robotic riot police capable of spraying thousands of bullets per second to squash potential riots. Increased surveillance and control of communication networks would make it possible to preempt planned demonstrations. It is also possible that a number of prison functions might be handled by robotics to make it cheaper to incarcerate more of the population. As noted in chapter 2, the United States at present has 5 percent of the world's population and 25 percent of the world's prison population. Robotics might make it possible to double or even triple prison space, maintaining an even larger prison population at a more cost-effective level. It may not be an unrealistic national goal to try to double or quadruple the present prison population while keeping costs flat simply by using robotic prison guards.

For most readers, the absurdity of the previous paragraph should be evident. For the rest, I should point out that a large prison population is almost always going to be one of the worst options for consequentialists. It merely treats the symptoms, not the disease, and does so in a particularly brutal way. Even if the prediction is good about how the United States will handle the situation, it is not one that a consequentialist should seriously entertain.

More plausibly, it might be thought that extending the current welfare system is a much better idea than something as drastic as BIG. Certainly this would be a serious improvement over increasing the brutality of the police state, but it still would not be nearly as effective as BIG in raising aggregate utility. Consider first the problem of unemployment. The difficulty people have in thinking about unemployment is surprising. At the height of the recession a few years ago, I asked my students a question after first explaining a situation thus: Suppose you have a friend who you see day after day lying on a couch watching TV. Your friend has no job. You say, "You ought to get a job." There are two ways this phrase can be understood: "It would be in your interest to get a job (so you don't waste your life watching TV)," or, "You have a moral obligation to get a job (rather than being a parasite on society)." In which sense do you mean this? The answer was that it was in both senses. Most of my students think it would be in their friend's interest to get a job and that he has a moral obligation to get a job.

To see the problem here, we need two further points. The first of which is that there is nothing special about the TV watcher. The same thinking applies to almost all the other unemployed (some may have special circumstances like a physical or mental disability; we can safely ignore this complication for present purposes). So at the height of the recession, this same imperative to get a job would have applied to 10 percent of the population.

The second point is philosophical: that "ought" implies "can." What this means is that if I say you ought to do X, then this implies that you can do X. Suppose I say you ought to save two drowning swimmers 5 km apart. You say there is a small chance and a great risk to your own personal safety that you might be able to save *one* of them but there is no way you can save both. And that by the time you get to one victim, the other will surely be drowned. We can see that there is no meaningful sense to saying you ought to save both when you cannot possibly save both.

But now we can see that the same point applies in times of high unemployment. When the number of jobs available greatly exceeds the number of unemployed, it makes no sense to say that everyone ought to get a job. There is a composition fallacy at work. It may be true that each runner might win the race, but it does not follow that every runner can win the race. Similarly, even if each person could get a job, it does not follow that every person can get a job. We must keep this in mind when we think through the proposal: it is utter nonsense to think that everyone ought to get a job.

We could, of course, consistently demand that everyone *try* to get a job, knowing full well that most will fail in this due to robotics displacing workers. But what would be the point of such an exercise? It would probably be a more useful and less soul crushing exercise to demand that the unemployed dig holes on even days and fill them back on odd days. Yes, the task would be as big a waste of time as looking for a job in times of high unemployment, but at least it would be good exercise. People who give up looking for work during periods of high unemployment ought to be treated as national heroes. If the economy is not producing enough jobs, then those who give up are making it possible for others to find work.

In times of high unemployment, it would make sense to decouple an "actively seeking work" stipulation for welfare benefits. Furthermore, we would not want welfare recipients being actively discouraged from seeking work. Welfare should encourage welfare recipients to work when it is available, so part-time or seasonal work ought to be encouraged, not discouraged. And if some have a business idea that they want to pursue, even better. To do this, there should be some means to allow such work to supplement welfare rather than being forced to choose between, say, low paying temporary work and welfare. Such a choice would discourage people from reentering the work force.

Once we stipulate that there ought to be no work requirement and that work ought to be encouraged by allowing people to supplement rather than replace their welfare payments when they have low wage jobs, then we can see that there would be few differences between such a modified welfare system and BIG. And to the extent that there are differences, the advantages are on the side of BIG. The similarities are clear: both the modified welfare system and BIG have no "actively looking for work" requirement, and both would not take away payments if any work is found.

In terms of advantages, one is efficiency.[30] It takes a small army of government officials to run the nearly 200 welfare programs in the United States. BIG would eliminate most of these because of its uniform nature: everyone receives BIG. There is no need to hire, sometimes Gestapo-like, officials to make sure that people are looking for work or not double-dipping: working and also collecting welfare or working part-time and collecting more welfare than entitled.

The second reason to prefer BIG is that it will reduce the "two citizen" view implicit in the welfare state model. The two-citizen view says there are two types of citizens: citizens proper and those who live off welfare. As argued in chapter 3, the two-citizens view is false. State capital justifies a BIG income. It is only false consciousness that drives this ideology. As I will explain in the following chapter, the two-citizens view is also harmful to aggregate happiness.

Another possibility would be to reduce the workweek to, in effect, divide up those remaining jobs. So for example, if the workweek was reduced from 40 to 20 hours, then business would have a large incentive to hire more workers. I will say more about this strategy in a later chapter and so will confine myself to a few brief remarks. First, reducing the workweek speaks only to the demand side of the problem. As robots take away certain jobs that are easily emulated by robots, for example, the ability to drive or dispense anesthesia, the residual demand for human labor will be for certain human intellectual abilities that are harder to emulate with robotics, for example, mystery novel writer or marketing executive. The skills to perform such jobs are not uniformly distributed in the human population: one needs some combination of natural endowment and education. Even if everyone has a natural endowment, it will take considerable time to reeducate workers for these jobs. Cutting the workweek does nothing to address the supply side, which involves keeping workers from starving while they retrain. Second, these are not mutually exclusive strategies. I contend that it will be necessary to do both: offer BIG and reduce the workweek to maximize aggregate well-being. Neither one implemented on its own will be sufficient.

Conclusion

The robotic revolution is already under way, even though it has not made a dramatic impact on the labor market at present. The argument of this chapter is that the robotic revolution will cause significant turmoil in the labor market. If there is ever a return to full employment, it will be after the result of major readjustments by workers and businesses. There is a significant chance that such turmoil, unless it is proactively addressed, will result in angry and frustrated workers.

In closing this chapter, it is worth repeating two points made earlier: BIG is proposed as a floor-level proposal rather than a ceiling-level one. As noted, it may be that we must cut the workweek in addition to BIG. It may also be necessary to offer citizens additional education and training in order to live in the new economy. So, to say that BIG is not enough is not a criticism of the present proposal. It should also be noted that it is not sufficient to merely say that BIG may have some bad consequences. Every public policy has negative consequences of some sort. The question for consequentialists, however, is whether there is a better overall net balance of good versus bad consequences for any given policy recommendation. So, for example, implementing BIG might have negative consequences for welfare officers who are charged with establishing eligibility for welfare payments. It is unfortunate that these people will lose a decent-paying job, but the upside is that more people will benefit under BIG. As noted, a successful criticism of BIG requires formulating an alternate policy that has a higher overall net utility: the standard is not a policy with no negative consequences.

Second, BIG is actually a conservative option, given our uncertainty about the future. There is near-universal agreement among experts who have examined the question of robots assuming many jobs presently done by humans. The residual disagreement is whether or not such displacement of human workers will result in massive unemployment. If, as I have argued, this displacement leads to high levels of unemployment, then the need for BIG to maintain peace is clear. The opposition here are optimists who believe that the economy will generate sufficient jobs for humans. As I argued above, optimists will also be committed to the view that there will be a massive expansion of the productive capacity of the economy to generate full employment. Under the optimists' scenario, the need for BIG to maintain peace will be greatly reduced, but then the cost of BIG, as a percentage of the economy, will also be relatively small. So, either BIG becomes necessary to keep the peace under massive unemployment, or the cost of BIG as an insurance policy becomes relatively small and easy to implement.

CHAPTER 6

BIG Happiness

Introduction

The academic study of happiness has undergone a revolution in the past 20 years or so. The work of psychologists, economists, sociologists, and other scientists today regularly make news headlines as the latest scientific investigations into happiness are brought to the public's attention. Most scientific interest has been focused on the "determinants" of happiness: what things make people happier or unhappier. A partial list of possible determinants studied by scientists include genomes, education, health, illness, unemployment, marital status, friendships, altruistic behavior, age, sex, race, TV watching, and income.

Some of the research from this scientific revolution has direct implications for consequentialist thinking about public policy, and BIG in particular. In order to provide a framework for appropriating this research, in this chapter we will assume a version of consequentialism: utilitarianism. Utilitarianism claims that the only thing that is ultimately good is happiness. Furthermore, we will assume a version of utilitarianism that says the right course of action for individuals and societies is the one that maximizes aggregate happiness. In thinking about policies that govern our society, we want to make sure that the policies adopted actually have the intended effect. So, utilitarianism is very interested in obtaining reliable predictions about possible consequences of different policy choices. The scientific study of the determinants of happiness is relevant because it can help us figure out which course of action is likely to generate the greatest amount of happiness. If scientists tell us education causes an increase in happiness, then as good utilitarians (other things being equal), we ought to fashion policies that promote education. If scientists tell us that watching TV, not exercising, and taking drugs promotes happiness, then (other things being equal), we ought to fashion policies that promote TV watching, not exercising, and taking drugs.

The work of social scientists supports the contention long made by utilitarians for redistribution of wealth and income. Taxing an additional $10,000 per year from someone like Bill Gates, who often makes over a $1,000,000,000 per year, would have a negligible effect on Gates' happiness, while giving a homeless person an additional $10,000 per year could have a dramatic positive effect on the homeless person's happiness. Notice that the argument does not require any assumptions about the economy growing: if we take $10,000 from Bill Gates and give it to a homeless person, the amount of money in the United States will remain exactly the same. To some, the argument seems almost magical. How can we increase happiness without increasing the total amount of money in the system? As we shall see, utilitarians answer this question with the idea of diminishing marginal utility of money.

This chapter updates the argument by discussing contemporary understandings of "happiness" and looking at empirical research, which investigates the relationship between money and happiness. These new developments, I shall argue, reinforce the utilitarian argument for distribution along the lines of BIG.

Defining "Happiness"

Social scientists differ on how to understand the meaning and measurement of the term "happiness." Although these different understandings are supportive of BIG as a policy, it will help us to understand the nature of this disagreement. We will start with some commonly made philosophical distinctions and work our way toward some controversies in social science.

We said that utilitarianism is a version of consequentialism and consequentialists seek to promote the good. What does "the good" mean in this connection? Answer: the good is whatever makes lives go well.[1] So, utilitarians claim that we ought to promote, as far as possible, whatever it is that makes lives go well. The question of what it means for a life to go well is the question of well-being, or prudential value.

Utilitarians claim that well-being is to be identified with happiness, that is, happiness is the sole prudential value. So the best version of any society is one that is organized in such a way that people are as happy as possible. Other things society might value, like education, good health, and wealth, are valued by utilitarians *only* to the extent that they promote happiness. So, a utilitarian would say that if wealth does not make you happy, then it is of no value to you. A little technical vocabulary should help us keep the distinction in mind. Intrinsic values are values that are valuable in and of themselves, while instrumental values are valuable only to the extent that they lead to intrinsic values. Utilitarians believe that there is only one intrinsic value: happiness. Utilitarians are not against education, health, and wealth by any stretch of imagination, but they are committed to saying that these things are valuable only to the extent that they lead to greater happiness. In other words, for utilitarians, education, health, and wealth are, at best, instrumentally valuable.

We can summarize this by saying that utilitarianism incorporates hedonism. Hedonism claims that pleasure or happiness is the sole prudential value. This may sound as if utilitarianism and hedonism are the same thing, but utilitarianism goes beyond hedonism: it tells us what to do about the prudential value of happiness, namely, act in such a way as to maximize it for all. One contrast here is an egoistic version of hedonism. It agrees that the sole prudential value is happiness, but it exhorts us to pursue our own happiness, not the happiness of all.

The main competitor to hedonism in the history of philosophy is perfectionism. Perfectionism is the view that the sole intrinsic good is the development of human excellences, which may include knowledge, rationality, friendship, creativity, autonomy, health, and physical prowess.[2] Typically, perfectionists value happiness only instrumentally. If happiness leads to greater health, for example, then happiness is instrumentally valuable. In the next chapter, I will argue for a hybrid view that combines both hedonistic and perfectionist elements. For the moment we will deal with only hedonism, and so focus on happiness.

In thinking about the meaning of "happiness," we should distinguish between the two senses of the word "happy." In one sense, it is merely a synonym for well-being in just the way that "hold"' is a synonym for "grip" and vice versa. In another sense, happy refers to a psychological state of some sort. As a first stab, we might think of happy in the psychological sense as the opposite of despondent or depressed. Indeed, the parallel runs deeper: just as we might think that psychologists have a professional interest in the mental state of depression, so too will they have a professional interest in happiness as a mental state. This interest, as noted, has blossomed in the science of psychology in the past 20 years or so.

Affective Theories of Happiness

As a rough taxonomy, we can think of the analyses of the mental state of happiness falling into two broad categories: affective and cognitive. The emotional state view of happiness is the major affective view of happiness in the social science literature[3] and has been defended by philosopher Daniel Haybron.[4] So, on the emotional state view, happiness is understood as the antipode to negative moods and emotions such as sadness, anxiety, and anger, frequently studied in psychology.[5] Happiness is to be identified with positive emotions such as "joy, interest, and pride."[6]

Thus, to be happy on the emotional state view is to have some favorable balance of positive moods and emotions over negative moods and emotions. To be unhappy is to have some unfavorable balance of negative moods and emotions over positive moods and emotions.

A competitor affective theory is the sensory pleasure view of happiness: happiness is the experience of sensory pleasure, with paradigmatic instances being sexual and gustatory pleasures. As Haybron notes, it may be that some of the classical hedonists, such as J. S. Mill, have not fully distinguished sensory pleasure from emotional state theory.[7] However, it seems there are good reasons to think the two as distinct. It is possible to experience sensory pleasure but not be in a positive mood. Imagine you have sex with your spouse at the regularly appointed time despite not being in a good mood: you suspect he has been cheating on you. You experience sensory pleasure during orgasm, but this in itself does not put you in a good mood. Of course, sensory pleasure and pain may often influence our moods: a delicious dinner may buoy our spirits and chronic or intense pain can negatively influence moods and emotions. So, sensory pleasure and positive moods and emotions are conceptually distinct, even if often causally connected.

While I believe that sensory pleasure is part of our understanding of happiness we will drop the question of whether sensory hedonism is part of happiness, as there is little relevant literature on the relationship between income and sensory pleasure.[8]

Cognitive Accounts of Happiness

In this section, we will look at cognitive accounts of happiness. Cognitive theories tend to be variants of the "whole life satisfaction view."[9] The basic idea, as its name suggests, is one in which individuals judge the overall quality of their lives as favorable.[10] Ed Diener and his colleagues formulated a comprehensive tool to investigate this sense of happiness.

The Satisfaction with Life Scale

DIRECTIONS: Below are five statements with which you may agree or disagree. Using the 1–7 scale below, indicate your agreement with each item by placing the appropriate number in the line preceding that item. Please be open and honest in your responding.

1 = Strongly Disagree
2= D isagree
3 = Slightly Disagree
4 = Neither Agree or Disagree
5 = Slightly Agree
6 = Agree
7 = Strongly Agree

______1. In most ways my life is close to my ideal.
______2. The conditions of my life are excellent.
______3. I am satisfied with life.
______4. So far I have gotten the important things I want in life.
______5. If I could live my life over, I would change almost nothing.[11]

Happiness, on this view, is to have a reasoned positive global judgment about the course of your life. It is important to understand that whole life satisfaction is not used to indicate whether someone is happy or unhappy, but, on this view, happiness *is* whole life satisfaction.

A number of philosophers have recommended the life satisfaction view.[12] Wayne Sumner claims happiness is

> a positive evaluation of the conditions of your life, a judgment that, at least on balance, it measures up favourably against your standards or expectations. This evaluation [...] represents an affirmation or endorsement of (some or all of) the conditions or circumstances of your life, a judgement that, on balance and taking everything into account, your life is going well for you.[13]

Let us note a few features about whole life satisfaction accounts. The basic unit of analysis is a propositional attitude. Propositional attitudes have an object (a proposition that describes some state of affairs) and an attitude (a mental state connecting a person to the proposition). Consider a proposition such as, "It will rain tomorrow." I might have various propositional *attitudes* to this proposition, for example, "I *believe* that it will rain tomorrow," or "I *desire* that it will rain tomorrow." In the case of life satisfaction accounts, the proposition is some global statement about the relationship between one's "standards or expectations" and one's life. The first item on Diener's survey nicely illustrates this: "In most ways my life is close to my ideal." The propositional attitude is one of (degrees of) agreement or disagreement. Thus, the happiest persons are those who would say, "I strongly agree that in most ways my life is close to my ideal." The unhappiest persons are those who would say, "I strongly disagree that in most ways my life is close to my ideal."[14] For whole life satisfaction accounts of happiness, the relevant domain of propositions are ones dealing with a person's life, or some important domain thereof, for example, one's work or family life.

Possible Evidence, Redistribution, and Aggregate Utility

The affective and cognitive accounts of happiness and their associated survey instruments in the social science seem to perpetuate the age-old debate about the primacy and importance of emotions versus rational elements of ourselves. Fortunately, as we shall see, we do not need to decide whether one theory, or perhaps both, provide a good account of happiness.[15] Both accounts point in the same direction: a distribution pattern of income and wealth that is more egalitarian will promote aggregate happiness. In this section, we will examine different possible relationships between income and happiness to give us some idea of what we should be looking for before turning to evidence from the social scientists.

The strongest case for redistribution in the name of aggregate happiness will come if the rich are made happier by being "unburdened" of their income and the poor are made happier by having more income, as illustrated in figure 6.1.

Second-best would be where income buys happiness for the less well-off but does not buy happiness for the rich, as in figure 6.2. Here, redistribution from the rich would benefit the poor and not diminish the happiness of the rich.

There are at least two ways that the evidence could go against the redistribution proposal if there is no correlation between money and happiness, or if the correlation is perfect. Figure 6.3 illustrates the first case. If figure 6.3 is correct, then redistributing money will have no effect on happiness since the rich and the poor are equally happy. (There could be other reasons for redistribution, but figure 6.3 would indicate that these reasons do not involve happiness.)

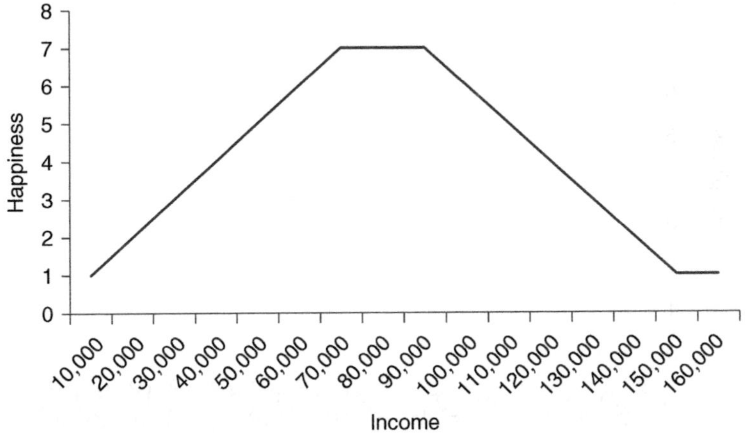

Figure 6.1 Income is a burden for the rich

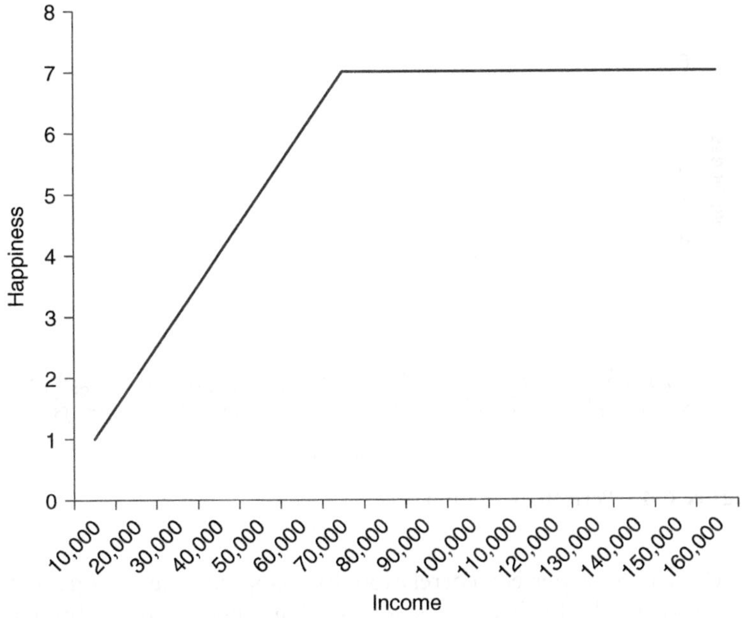

Figure 6.2 Income does not buy happiness for the rich

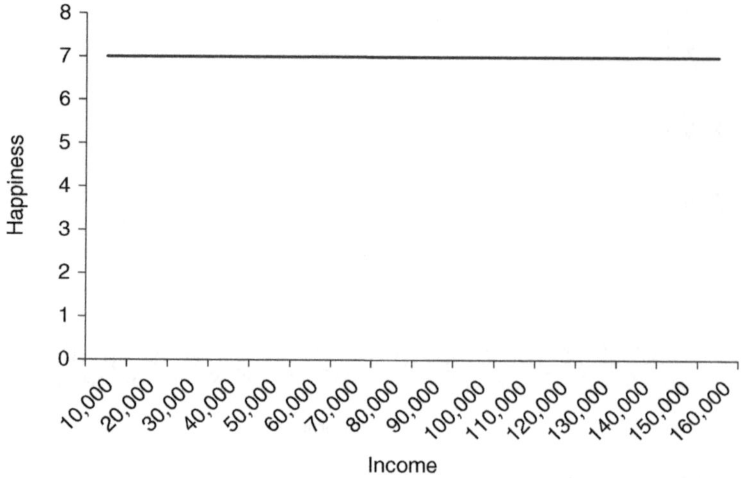

Figure 6.3 Income does not buy happiness

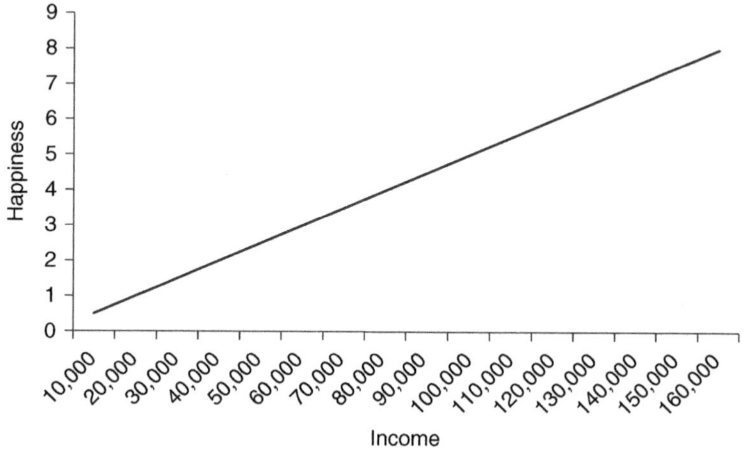

Figure 6.4 Perfectc orrelation

If there is a perfect correlation for happiness, as in figure 6.4, this too would be bad news for the happiness-from-nothing thesis. Figure 6.4 suggests that a one point gain in happiness is associated with an additional $20,000 in income. So increasing a person's income from $20,000 to $40,000 would increase their life satisfaction score from 1 to 2. But if $20,000 were taken say from someone earning $160,000, this would reduce his or her happiness by 1 point,

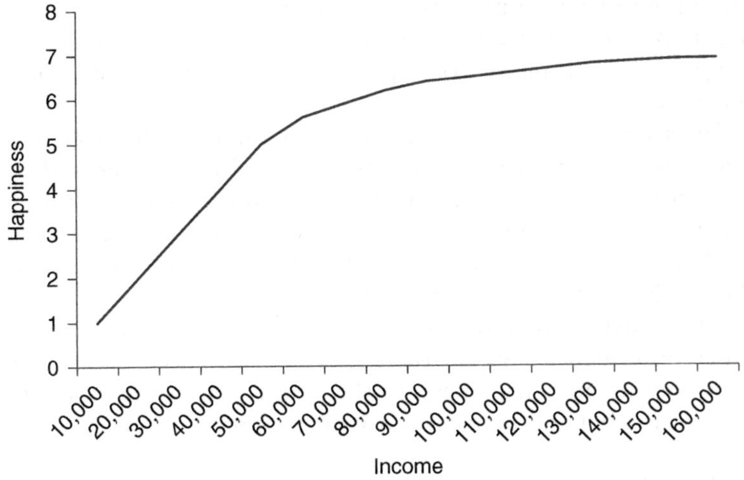

Figure 6.5 Diminishing marginal utility of income

so there would be no net gain in overall happiness. Of course, there might be other reasons to redistribute wealth, for example, that it provides a fairer distribution of happiness, but this would be bad news for the happiness-from-nothing thesis.

Not quite as good as our first two graphs, but a third best would be one where there is a diminishing marginal utility, as in figure 6.5. This would imply that money always buys happiness, but at a diminishing return at higher levels of income. So, unlike the first two graphs, it has the implication that taking away money from the wealthy does diminish their happiness, but only very slightly. The same dollar amount would dramatically improve the happiness of the poor. So, diminishing marginal utility of income is consistent with the happiness-from-nothing thesis.

Actual Evidence

As intimated above, the empirical evidence suggests that income affects the two components of happiness, affective and cognitive, differently. We will take these in turn.

Kahneman and Deaton have found a satiation point for income and positive affect at about the $75,000 income level.[16] Their data for positive affect is strikingly similar to our second-best case: money does not buy happiness for the rich (figure 6.2). The Kahneman and

Deaton result for positive affect has the implication that redistributing money from the rich to the poor would not affect the positive affect of the rich but it would positively boost the positive affect of the poor.

In other words, the Kahneman and Deaton result strongly supports the utilitarian case for redistribution. If income were redistributed from those making more than $75,000 to those making less, there would be no diminishing of positive affect at the higher income level and large gains for those at the bottom of the income scale, should the money be redistributed to them.

The evidence for cognitive understandings of happiness is notably different: evidence suggests there is no satiation point where money does not buy more happiness. In the past, this was often thought to be the case, but as Stevenson and Wolfers argue:

> While the idea that there is some critical level of income beyond which income no longer impacts well-being is intuitively appealing, it is at odds with the data. As we have shown, there is no major well-being dataset that supports this commonly made claim.[17]

If Stevenson and Wolfers are correct, then their results put to rest the idea that money does not buy happiness for the rich. However, their data also shows that a given amount of money buys more happiness for the poor than for the rich, that is, there is a diminishing marginal utility of income. Stevenson and Wolfers note this implication of the diminishing marginal utility of income: "going from $1,000 to $2,000 raises satisfaction by twice as much as going from $2,000 to $3,000 and by the same amount as going from $10,000 to $20,000."[18]

Thus, the Wolfers and Stevenson data strongly supports the utilitarian case for redistribution. For example, it says that raising the household income of the Joneses from $64,000 to $128,000 would have the same effect on their happiness as raising the income of their poor neighbors, the Smiths, from $4,000 to $8,000. Conversely, it tells us that if the Joneses make $128,000 and we redistribute $64,000 and boosted 16 families to $8,000 from $4,000, we would generate 16 times the amount of happiness that the Joneses lost through redistribution.

A look at some data from a Gallup Poll (December 6–9, 2007) will help provide further insight into the possible effect of BIG.[19] The Gallup survey has a four category inventory for life satisfaction: "very satisfied," "somewhat satisfied," "somewhat dissatisfied," and "very dissatisfied." In households with less than $10,000 annual income, only 24 percent rated themselves as "very satisfied," while 32 percent rated themselves "very dissatisfied." There is quite a striking difference compared with households with incomes in the $10,000 to $20,000 range. In this range, 47 percent of respondents indicated that they were "very satisfied" with their lives and only 11 percent rated themselves as "very dissatisfied." In other words, in the $10,000 to $20,000 category, twice as many were "very satisfied" and only a third as many were "very dissatisfied" as compared with the less than $10,000 category.

A somewhat naive prediction is that the lowest income group at present, where only about a quarter rate themselves as "very satisfied," would almost double with the implementation of BIG to a point where almost half would rate themselves as "very satisfied." I say it is naive because there are a number of confounding factors that almost always intrude when thinking about policy changes. To illustrate: the categories that the Gallup uses for income have quite large intervals. A BIG of $10,000 would push some just over the threshold into this category. If there is significant variation within the category of $10,000 to $20,000, then the results might be skewed. For example, if most of the gains in happiness in the second category occur with those closer to $20,000, then the boost might be less than predicted. Also, if social comparison accounts for a large portion of the variation in satisfaction ratings, then the effect on happiness might not be as great as the naive prediction. Consider that pre-BIG, households with no income have the lowest income. With a BIG of $10,000, these same households will have increased in their income in absolute terms, but in terms of social comparison, they still have the lowest income. On the other hand, it may be that BIG provides a bigger boost than the naive prediction indicates. It is no secret that many low paying jobs provide unpredictable employment. Part-time jobs and seasonal jobs typically offer workers very little guarantee in terms of regular employment. This bourgeoning segment of the labor market has been dubbed the "precariat" by Guy Standing.[20] The enormous uncertainty in the lives of the precariat worker is no doubt a source of unhappiness. Even a modest BIG of $10,000 could boost happiness more than we would predict on the basis of income alone, if the fact that it is guaranteed makes a difference to how vulnerable workers feel. Universal health insurance is also another confounding variable. We might expect that some of the unhappiness of the lowest income group is due to lack of affordable health care. If BIG and universal health care were implemented at the same time, it would be hard to predict the level of happiness for the least well-off. Still, there is every reason to predict that the least well-off will gain significantly in terms of happiness.

The Gallup Poll also shows how little effect decreased income would have for the wealthy. The second highest income category in the Gallup survey is household income between $250,000 and $500,000. Imagine if the proposed VAT of 14 percent is introduced. Potentially some people in this category (those making $500,000) will pay approximately an additional $40,000 to $50,000 in tax.[21] Even this hefty increased tax bill would not be sufficient to move them down to the third highest income category of $150,000 to $250,000. But if it did move some down to the third highest category, the effect on life satisfaction would be extremely small. In the second highest income category ($250,000 to $500,000), 93 percent rated themselves as "very satisfied" while 90 percent in the third highest category ($150,000 to $250,000) rated themselves as "very satisfied." Since there is little difference between the two categories, we can see that it is a safe prediction that a reduced income will only have a very small effect on the cognitive dimension of happiness.

It is important to underscore that the thrust of the utilitarian case for redistribution is that it creates happiness-from-nothing. We mentioned this earlier, but it is worth investigating further.

To see what is at issue, imagine the Utilitarian fairy asks us to choose between raising a poor person's happiness by one unit or a rich person's happiness by one unit. In the case of the former, this means doubling the income of someone earning $10,000 per year to $20,000 per year. In the latter case, it means increasing the rich person's income from $100,000,000 to $200,000,000 per year. We might think this is implausible. True, $100,000,000 more per year is a significant increase, but it also seems reasonable that someone who is making $100,000,000 has far fewer monetary impediments to happiness already. On the other hand, we can see why $10,000 in extra income could make a significant difference to someone making only $10,000 per year. The research from Stevenson and Wolfers indicates that the increase in happiness for both would be the same. So, in terms of promoting aggregate utility of just these two individuals (thus ignoring any effects on others in the economy), it would make no difference whether the rich person's income was increased by $100,000,000 or the poor person's by $10,000. In either case, aggregate utility would be increased by one point.

To bring the point home to BIG, we can also see that taking half the income from someone earning $2,000,000 a year will reduce his or her happiness by X, but redistributing in the manner described would add 100X, so the next gain would be 99X. It is this almost magical creation of happiness that pushes utilitarians to income redistribution. Without increasing the size of the economy one iota, we can increase gross national happiness (GNH).

It may help here to contrast this thinking with the way economists long thought about the relationship between happiness and economic output, which was that gross domestic product (GDP) was a reasonable proxy for the average happiness of a nation.[22] On this thinking, the way to improve GNH is to increase GDP. The evidence cited in this chapter supports this claim, at least for the cognitive understanding of "happiness." However, we have also shown that even if the GDP remains fixed, a better distribution of income will increase GNH. Happiness from nothing, then, means that we can increase GNH without increasing GDP.

Measuring Happiness

The argument depends crucially on insights gained from social science about the relationship between happiness and income. We will consider three different doubts about this relationship, and therefore three different challenges to the argument. The first is that we should not trust data from social sciences because social scientists use mostly survey data. Worries about survey data include the idea that people may lie, there may be systematic biases, and that happiness is too personal to ever be measured.

This is certainly not the place to mount a full-scale defense of happiness surveys, so I will confine myself to a few brief points. Like any scientists, social scientists put enormous work into investigating their scientific instruments. In particular, social scientists typically ask about the reliability and validity of their instruments. Reliability refers to the idea that the instrument will give consistent readings of the phenomenon under investigation. Validity refers to the idea that one is measuring what one in fact intends to measure. For example, suppose we measure the height of some seedlings. On day one they measure 3 inches, on day two they measure 2 inches, and on day three they measure 3 inches. Our results suggest that the seedlings shrunk and grew radically in the course of a couple of days. We then find out that our measuring tape is made of cotton and it stretches when wet. On day two it rained and so our tape measure made it appear that the plants shrunk. In this case our tape measure is unreliable. Suppose then we switch to using a metal tape measure that is not affected by moisture. If we get a consistent reading of 4 inches on days one, two, and three, then we have good reason to think the tape measure is reliable. If, unbeknownst to us, the first inch of the tape measure is missing, we would get a reliable but invalid measure of the height of each plant.

Social scientists have put enormous effort into investigating the reliability and validity of their surveys.[23] To check reliability, social scientists may survey the same population at different times. If there is a radical difference in average happiness and no other explanation for the variance, then this is evidence that the survey is not reliable. To check the validity of their measures, happiness scientists will look for independent measures of happiness. For example, among your friends and acquaintances you know some who are happier and some who are less so. Social scientists will record these estimates and compare them with self-reports. So suppose you are asked about two friends, Jack and Jill. You rank Jack 9/10 on the happiness scale and Jill 4/10. Scientists will then compare these numbers with Jack and Jill's self-reports of happiness. Scientists have found that these numbers correlate to a high degree, suggesting that surveys that rely on self-reports of happiness are measuring what they intend to measure.[24]

As mentioned, this is not the place to provide a defense of the science of happiness. True, if there are good reasons to doubt this science then there is reason to doubt the idea that BIG will promote aggregate utility. But skepticism about the science of happiness works both ways. Some may say that there is no reason to think that BIG will increase happiness, and so this undermines one reason to endorse BIG. Of course, this also undermines a reason to resist BIG: the idea that redistribution will make people unhappy. That is, an opponent of the argument cannot consistently say, "there is no reason to suppose BIG will increase happiness, because the science of happiness is illegitimate," and "we should not adopt BIG because higher taxes will make people unhappy." The problem for this line of objection, in other words, is that if there are doubts about measuring happiness, then this cuts equally well against the idea that increased taxes affect the happiness of the rich.

Happiness and Income: The Easterlin Paradox

A less extreme form of skepticism agrees that we can measure happiness, but suggests that the empirical data shows that money does not buy happiness. Support for such a claim may appear to be suggested by a seminal article published by Richard Easterlin in 1974,

"Does Economic Growth Improve the Human Lot? Some Empirical Evidence." This is the catalyst for much of the interest and growth of happiness studies by economists and other social scientists.[25] Easterlin's "paradoxical" conclusion has sometimes been summarized as "money doesn't buy happiness," but this is far from what Easterlin said.[26] It is worth attending to exactly what Easterlin concluded:

> The concern of this paper has been with the association of income and happiness. The basic data consists of statements by individuals on their subjective happiness, as reported in thirty counties, including eleven in Asia, Africa and Latin America. Within countries there is a noticeable positive association between income and happiness—in every survey, those in the highest status group were happier, on the average, than those in the lowest status group. However, whether any such positive association exists among countries at a given time is uncertain. Certainly, the happiness differences between the rich and the poor countries that one might expect on the basis of within-country differences by economic status are not borne out by the international data. Similarly, in the one national time series studied, that for the United States since 1946, higher income was not systematically accompanied by greater happiness.[27]

One important thing to notice is that Easterlin says there is a positive correlation between income and measures of well-being within a country. This result, at least with respect to life satisfaction, has been repeatedly confirmed, including the recent results noted above by Stevenson and Wolfers: "there is no evidence that the slope flattens out beyond any particular 'satiation point' in any nation."[28] So, Easterlin's findings support the following conclusion:

1. There is a diminishing marginal utility of income within countries at a particular time.
2. There is no correlation between income and happiness between nations.
3. There is no correlation between income and happiness between time periods in a nation.

Given the second and third conclusion, it is easy to see how some might think that money does not buy happiness. After all, if rich nations are not systematically happier than poor nations, and if nations do not get happier as they get richer, then surely there is room to doubt whether income causes happiness.

In doubting that money is correlated with happiness, some commentators seem to have lost sight of exactly what is paradoxical about Easterlin's result. If Easterlin found no correlation between income and happiness, then 2 and 3 would be easy to explain. The trouble is that 1 indicates a robust correlation between happiness and income, whereas 2 and 3 do not.

Most of the debate between Easterlin and his critics turns on disputes about 2 and 3, that is, the between-country comparisons of happiness and the time series comparisons within a country. As I hope to show, the debate is not really relevant for our purposes, so in discussing the Easterlin Paradox, I shall simplify the matter somewhat. Easterlin's results seem to challenge the "absolute" view of income. The absolute hypothesis says that income affects happiness irrespective of what has happened in the past and irrespective of how one's peers are doing. The relative hypothesis says that the effect of income on happiness is influenced by history or one's peers. There are two influential versions of the relative hypothesis. The adaption theory predicts that changes in income tend to have relatively short-term effects on happiness. A raise at work makes us happier, but after several months or years, our happiness level will tend to settle back where it started. The social comparison theory says that the effect of income on happiness is directly proportional to how our peers are doing. If I get a $5,000 raise while my peers get a $10,000 raise, I may not be happier even though my income has risen in absolute terms.

Looking at the between-country comparisons, if the absolute conception is correct, then countries with higher average income should have higher life satisfaction ratings than poorer countries. This is indeed what Stevenson and Wolfers found:

> That citizens of rich countries have higher well-being on average than citizens of poor countries. The correlation between average well-being and log GDP per capita in 2010 is 0.74; not only is there a clear connection between well-being and GDP, but GDP in fact explains (in a statistical sense) more than half of the cross-national variation in well-being.[29]

Similarly, if the absolute conception is correct, then rising national incomes should be correlated with rising life satisfaction ratings. This is what Sacks, Stevenson, and Wolfers found: a correlation between growing economies and higher life satisfaction ratings.[30] This suggests that absolute, not relative, income is the more plausible hypothesis, for if the adaption version of the relative theory is correct, we should expect no growth in life satisfaction scores. The data from Sacks, Stevenson, and Wolfers, however, show an upward trajectory of life satisfaction scores with growing economies.

My point here is not to suggest that Easterlin is wrong, only to illustrate the controversy over 2 and 3. We are now in a position to see that it does not matter whether Easterlin or his critics are correct for the proposal under consideration. It will help to turn points 1, 2, and 3 into questions that we might ask as utilitarians:

Q1. Should we redistribute income within countries at a particular time?
Q2. Should we redistribute money between nations?
Q3. Should we aim to increase the total income of a nation?

If Easterlin's critics are correct, then the answers, from a utilitarian perspective, are "yes." Since money always buys happiness (at a diminishing rate), it would make sense to redistribute money within society (other things being equal), so a positive answer to Q1 is mandated. It would also make sense to redistribute money from rich to poor nations (other things being equal), since the happiness of poor nations would go up more than the happiness of rich nations would decrease, so an affirmative answer to Q2 is mandated. And finally, if happiness always increases with total income, then (other things being equal), nations should aim to grow their income, and so a yes to Q3 is mandated.

If his critics are wrong and Easterlin is correct, then the answer to Q2 and Q3 is "no." If Easterlin is correct, there is no correlation[31] between happiness and income between nations, and there is no correlation between the increasing income of nations and GNH. Hence, there would be no utilitarian reason to redistribute money between nations or attempt to increase GDP. Interestingly, if Easterlin is correct, then there is reason to redistribute income within a country at a given time period, since higher income is associated with increased happiness. Thus, an affirmative answer to Q1 is mandated from a utilitarian perspective whether Easterlin or his critics are correct.

We have, then, another means to illustrate what is so paradoxical about Easterlin's results. On the face of it, it seems that the same answer should be given to all three questions. If there is no relationship between income and happiness, then the answer to Q1, Q2, and Q3 should be "no." If there is a relationship between income and happiness, then the answer to Q1, Q2, and Q3 should be "yes." Easterlin's result suggest an inconsistent response: sometimes "yes" and sometimes "no."

In any event, the important point for us is that the question of whether to redistribute income in the United States in the beginning of the twenty-first century clearly falls under the scope of Q1. As we have noted, both Easterlin and his critics would agree that the utilitarian should redistribute (other things being equal). So, we do not need to sort out this controversy to see that empirical research supports the case for BIG.

Causation and Happiness

We noted above that most social science data works with correlational data between income and happiness; very little addresses the issue of causality. One way to sort out the issue of causality is with longitudinal studies. Longitudinal studies, as the name suggests, typically sample a population over time. If we have two correlated variables X and Y, and we are not sure which is the cause and which is the effect, we look to see whether one precedes the other in time. If X causes Y, then X should be present earlier than Y. Conversely, if Y causes X, then Y should be apparent earlier than X. To illustrate, consider a study by Ed Diener and his colleagues.[32] The study assessed the cheerfulness of college freshman and their income 19 years later. The group assessed as more cheerful in college earned more than their less cheerful cohorts 19 years later. The evidence here is suggestive that happiness causes increased earnings. As with any attribution of causality, there are significant worries about confounding variables. In this case, an obvious one is parental income. It does not seem out of the realm of possibility that freshmen whose parents earn more may be happier, and higher parental income may explain why their children are able to earn more. To sort out possible confounding influences, researchers "control" for such variables. In this study, when parental income was "high," defined as above $50,000, the most cheerful college students made a whopping $25,000 more than their less cheerful counterparts, no matter whether they had high income or low income parents.

What are the implications of this for the happiness-for-nothing hypothesis? In short, not a whole lot. For a start, these findings do not overturn the view that income causes happiness. The research is consistent and supportive of a bidirectional causal direction: happiness causes income and income causes happiness.[33] Similar bidirectional effects have been observed for other determinants of happiness: health can cause happiness and happiness can cause health, marriage can cause happiness and happiness can cause marriage.[34] In terms of policy, if a policy makes a population happier, then it is likely to make the population healthier. Conversely, if a policy makes the population healthier, it is likely to make the population happier. Thus, the fact that happiness increases income is no reason to think that income does not increase happiness.

There is no suggestion here that BIG is the only policy or the single best policy for raising aggregate utility. Numerous studies confirm that most of the variance in happiness levels in developed nations are accounted for by genetic differences. Typically, estimates suggest that 40 to 80 percent of the variance is due to genes, with the higher number probably closer to the truth.[35] In other work, I have argued that we could leverage the fact that genes contribute so much to happiness and develop a pill to increase our happiness for an amount of ten billion or so. In strictly economic terms, such a onetime development cost is small with respect to total GDP. So, if a utilitarian had to choose between BIG and happy-people-pills, then the latter would win, hands down. Of course the choice is a false dilemma, as both strategies would serve utilitarian aims of increasing aggregate utility.

Note, too, that these remarks apply to the population as a whole, we would do more to raise total utility if we adopted the happy-people-pill proposal. On an individual level, these generalizations may not be true. For example, a penniless homeless person would probably benefit more from BIG than happy-people-pills. Fortunately, as noted, we need not decide between these policies: utilitarians ought to support both strategies.

Socialism as an Alternative to BIG

It may be thought that the argument does not go far enough. If we take the diminishing marginal utility of income phenomenon seriously, then any disparity of income seems to leave room for further redistribution in the name of aggregate utility. For taking any additional income from the rich should always affect their happiness less than giving it to the less well-off. The only time this fails to be the case is when everyone has the same income. In the United States, this would work out to an income of about $42,000 (in 2012).[36] The logic of the argument, in other words, seems to indicate that we ought to provide a BIG of $42,000. But this is absurd, it would be equivalent to the radical equality called for by some communists and others, so the argument is flawed.[37]

The objection brings out an important point: the argument assumes that other things will be equal. As consequentialists, we must think of policy as aiming at increasing aggregate happiness, and so any effects of a proposed policy themselves are part of the equation. In this case, there is an obvious question about motivation. If there is complete equality of income, for example, we may worry that not enough people will participate in the economy or that it may be difficult to fill dangerous or unpleasant jobs. Here is a job you probably do not want if you make the same as everyone else: the job involves putting on a pressurized diving suite and diving into city municipal sewage plants to make repairs. The work is extremely hazardous and unpleasant—to put it mildly. Who would take such a job if the alternative is to work for equal pay in more pleasant and benign conditions?

The point here is a familiar one: a society of strict equality faces a motivation question: how to motivate people to do unpleasant or dangerous work, or indeed any work at all. Of course, if few people work, or are willing to work in dangerous or unpleasant jobs, then we will all suffer. The rich and the poor alike use the municipal sewage system, and it will be a crappy outcome for all if no one is willing to put on a pressurized suit and go fix the plumbing. Until we hear a plausible theory as to how people will be motivated to work under strict equality, we should follow suit with Rawls and other liberals and accept that differences in income are an acceptable means to ensure that a society remains productive.

Conceding that strict equality will probably lower aggregate utility does not amount to the admission that any redistribution will lower aggregate utility. For example, taxing someone who makes 2 million a year at 90 percent would mean a tax bill of $1,800,000. As draconian as a 90 percent tax rate seems, in absolute terms, the person who nets $200,000 a year would still end up with an income about 5 times the national average. This may be sufficient motivation for the person in question to do an unpleasant or dangerous job, or to sell her special natural talents on the labor market. Ultimately, this is an empirical question: it is not one that can be decided by armchair speculation. How much difference in income is necessary to motivate people with special talents to perform dangerous or unpleasant tasks? Studies by psychologists and other social scientists are needed to help us answer this question. Such studies would have to look at what people do rather than what they say. People often claim they will work less if taxes go up, but there is no evidence that this is in fact the case.[38]

Note too that the question for consequentialists ultimately is not how much we can raise taxes before total productivity goes down, but *how much can we raise taxes before aggregate utility goes down*. In theory, GDP could shrink quite drastically, say by 10 percent, yet aggregate utility could still rise if income is distributed from the rich to the poor. The reason, of course, is that money buys more happiness for the poor than for the rich. My guess is that a much larger BIG payment is theoretically possible. For example, a 20 percent VAT might provide an income closer to $15,000 and this would do more for aggregate utility than the proposed $10,000 income. I am not against this in principle. As mentioned several times, the proposal here is a floor-level income rather than a ceiling-level. A higher BIG would face exponentially greater opposition. So, strategically, it would probably be best to start with something more modest and work for gains from there. An initial BIG of $10,000 would provide some experimental data from which to calculate effects on productivity and aggregate utility.

The Welfare State as an Alternative to BIG

We should also consider the welfare state as an alternative here. The welfare state could also buy into at least part of the argument: if the rich are still motivated to work with much higher taxes, then more money could be redistributed to those who pass the actively looking for work (ALW) requirement.

One cited advantage of ALW is that it provides a spur to look for work. The thought is that without ALW, people would tend to become lazy and stop looking for work. The usual cited disadvantage of receiving welfare and not looking for work is that it is a form of parasitism: welfare bums suck the life out of the hardworking tax payers. We will discuss this objection in chapter 8.

Another advantage of ALW is for the unemployed themselves: studies show that the unemployed are less happy than the employed, even when income is controlled for. So, for example, if you have a job paying $30,000 a year and you are laid off and go on unemployment insurance, which pays $12,000 per year, we should predict a drop in your happiness simply because your income has fallen. Empirical studies, however, show that an unemployed person's happiness drops more than what we would predict from the drop in income alone.[39]

So, if unemployment assistance with an ALW requirement brings benefits to the unemployed themselves, and if it provides the spur to action to get some of them back to work, then we should predict that their happiness will increase more than the associated increase in income.

The fact that the unemployed are unhappier than the employed even when controlling for income ought to strike us as very counterintuitive. After all, the unemployed have more leisure time than the employed. If we did not already have the empirical data in hand, it would seem a natural assumption to think the unemployed should be happier than their employed counterparts, controlling for income. Let us think of this as "the paradox of the unemployed." As consequentialists, we cannot ignore the data and it appears to provide at least some support for the welfare state with its ALW requirement.

In response, I want to say that what is sometimes cited as a feature of the welfare system is actually a bug. Implicit in the welfare state is what we might think of as the "two-citizens view." There are good citizens, citizens who work and contribute to the good of society. Good citizens have a duty to give over some of their earnings (perhaps according to the Rawlsian difference principle) to the less fortunate. Lesser citizens, the citizens who are the recipients of the redistribution, should move to the good citizens category as soon as possible, if they can.[40] One way to make sure they do is to have the ALW requirement. Welfare administrators are charged with making sure that as many people as possible move from the lesser citizen category to the good citizen category.

Just to be clear, I am not saying that the two-citizens view is explicitly endorsed in practice or by proponents of the welfare state (for example, Rawls). What I am saying is that it is at least an unintended consequence of the welfare system. Consider the mechanics of applying for welfare. Typically, this means going to a government office and starting the application process. Many people find this in itself embarrassing and/or humiliating.[41] Prostrate in the office, you will be subject to questioning and monitoring by public officials. Are you really in need? Are you looking for work? Can you prove any of this? The system is designed to ensure that there is no fraud: that good citizens, or those capable of being good citizens, are not masquerading as lesser citizens. At least some of the unpleasantness of the process is by design: getting welfare should not be easy or pleasant, which ensures that the ranks of lesser citizens never swell beyond absolute necessity.

Clearly this process takes away from aggregate utility, especially when unemployment is not caused by laziness but structural changes in the economy. With BIG, there is no need to go to a public official and declare one's change of citizenship status from good to lesser. BIG is automatically given to everyone: there is no need to bare one's self to inspection and probing by public officials. In terms of redistribution, BIG says we are all in the same boat: we all receive BIG.

Note that there is no claim that unemployment will not cause unhappiness with the advent of BIG. Rather, the claim is that a future where we have BIG, as opposed to welfare and unemployment, will cause less overall unhappiness for the unemployed, even when the payments to them remain the same. The conjecture is that a change to ones' social status and standing associated with the two-citizens view negatively affects one's happiness, and that this accounts for at least part of the paradox of the unemployed.

There is some empirical support for this conjecture. Areas with high unemployment are ones where the unemployed are happier. This should strike us as counterintuitive, as the paradox of the unemployed. For example, one might think that being laid off in an area with 10 percent unemployment should negatively affect one's happiness less than an area with 20 percent unemployment. After all, the chances of getting a job in an area with 10 percent unemployment are so much better than one with 20 percent unemployment. However, studies show that the unemployed are actually happier in the higher unemployment area compared with their counterparts in the lower unemployment area. This suggests that public perception plays a large role in explaining the paradox of the unemployed. In areas with high unemployment, the unemployed are not blamed for their situation as much as in areas of low unemployment. So, even in areas with high unemployment and less chance of securing a job, the unemployed are not as unhappy because they feel less responsibility and take less blame for their situation.[42]

This indicates a dilemma for the welfare state's two-citizens view and its attendant ALW requirement. If unemployment rates are high, then the ALW surveillance requirement is an extremely inefficient use of resources.[43] If unemployment rates skyrocket to 25 percent or higher, like during the Great Depression, the number of welfare workers would have to increase drastically. These officials would be charged with making sure that a quarter of the work force is looking for work when clearly there would not be nearly enough jobs for everyone. This would be to no good effect and a huge waste of resources. On the other hand, if the unemployment rate is lower, then perhaps there is more need for the ALW surveillance requirement, at least according to apologists for the welfare state. However, we know that the ill effects for the unemployed will be higher when the unemployment rate is lower. Part of this ill effect is a direct result of the two-citizens view, which is reinforced by the ALW surveillance requirement. By removing the ALW requirement, we can reduce some of the ill effects of unemployment. Indeed, where BIG is seen as a just dividend (as argued in chapter 3), rather than a handout from good citizens, then it is likely that the ill effects of unemployment will be even further mitigated.

Conclusion

The case for redistributing income is based on very robust data. As noted above, there are other things too that might be done to increase aggregate happiness, but income redistribution is certainly one.

Even with the solid reasons given, we must admit that ultimately there is an empirical element to the proposal. There are no cases of countrywide, long-term basic income experiments. It is possible, in principle at least, that a BIG experiment could reduce aggregate happiness. If the experiment were run in this country for ten years and aggregate happiness declined, and other factors were eliminated as possible causes for the decline, then utilitarians would have to come out against BIG. I raise this possibility only to indicate that the experiment is data driven. It does not amount to much of a concession. After all, it is also an empirical proposition that pigs will not start to fly right after the conclusion of the next Super Bowl. Theory tells us that it will not happen, that pigs will stay firmly on the ground, but if we saw them flying we would have to admit our theory was wrong. Similarly, if redistributing income along the lines of BIG results in lower aggregate utility, then we would have to admit our theory is wrong. In the meantime, it is safe to assume pigs will not fly and BIG will increase aggregate utility.

CHAPTER 7

BIG Freedom

Introduction

In the previous two chapters, it was argued that BIG provides a good means to ensure two social goods: peace and happiness. We anticipated a capitalist's objection to this line of argument in chapter 4, namely, that redistribution in the name of the social good is morally wrong. The argument was that capitalism itself requires reigning in of individual economic decisions in order to ensure the survival of capitalism itself. There is, however, still a powerful response open to the apologist for the sort of laissez-faire capitalist who wants to resist society's taking happiness as an important desideratum for distribution:

> Freedom is the highest social value. This was proven by Huxley in his novel Brave New World: it is a society that lacks freedom. The lives of the citizens of the Brave New World are carefully controlled by the political leaders of society. True, they are much happier than we are. However, no one in their right mind would trade our world for theirs. This shows that freedom is a much more important political value than happiness. Those who invite us to redistribute wealth and income in the name of happiness forget the terrible price of their proposal: freedom. Capitalism without redistribution in the name of the social good gives us the greatest freedom,[1] while reining in capitalism in the name of happiness reduces freedom. This is a cost we should not pay.

The utilitarian has an answer to this objection: freedom is valuable only to the extent that it promotes happiness. Thus, the utilitarian will disagree vehemently with the idea that freedom is more important than happiness. This line of defense is not one that can be endorsed here, since it will be argued later in the chapter that freedom has intrinsic value. As noted above, this means that we must reject utilitarianism.

Since the assumption is that freedom is an intrinsic value, the objection above on behalf of the laissez-faire capitalist is an important one. It suggests that a fundamental decision must be made: should we sacrifice freedom for the sake of happiness? If freedom is more important, then the argument of the previous chapter seems to be trumped by the point that distribution along the lines of BIG reduces freedom. If happiness is more important, then the argument for redistribution along the lines of the previous chapter is not thwarted. However, it does seem to require an argument that happiness is more important than freedom.

To argue that happiness is more important than freedom requires taking on many opponents. The rallying cry of freedom or liberty, from the Enlightenment on, is heard across the political spectrum. Thinkers such as Locke, Kant, Hegel, Marx, Green, Tawney, Rawls, Nozick, and Cohen have claimed freedom has a central role in political thinking.

Fortunately, at least for the purposes of this work, we do not need to decide whether happiness or freedom is of more important political value. The argument of this chapter focuses on freedom and parallels the diminishing marginal utility argument of the previous chapter. The basic idea is that redistribution of income from the rich to the poor in the form of BIG will increase gross national freedom (GNF). That is, redistribution along the lines of BIG creates "freedom from nothing" in the same way that redistribution of income creates "happiness from nothing": no additional economic output is necessary to create more happiness and freedom. The argument, in other words, is that unrestrained capitalism does not provide the maximum amount of freedom. The social values of happiness and freedom both support BIG.

The major complication in setting out this argument is to understand the term "freedom." The political debate over the past few centuries has invoked a number of understandings of the notion of freedom, and so there is much danger in invoking freedom without considerable explanation. To see our way into the labyrinth, we will start with some aspects of what we might think of as our "ordinary understanding" of the term "freedom." By our "ordinary understanding," I mean simply how the notion is used by those who do not specialize in political theory: how the proverbial man and woman on the street use the term. Part of this ordinary understanding is the idea that money and freedom are intimately linked, other things being equal. If you have more money, you have more freedom. We will illustrate this idea with an example of a couple who won the lottery a few years back. Later in the chapter we will contrast our ordinary sense of freedom with more specialized senses of freedom used in political philosophy.

Freedom and the Lottery

As just noted, it is a well-entrenched idea, at least in nonphilosophical contexts, that more money leads to more freedom. Consider the story of Jack and Debra Woodruff who won a million dollars in 2011 in the Hoosier Lottery. Of course they were very happy to win, especially since they initially checked the numbers and thought they had won $250,000, not a million. Although the couple was very happy, this was not the best thing according to Jack Woodruff:

> For Jack Woodruff, winning $1 million means one thing: freedom. Jack is a 28-year veteran of the Fort Wayne Police Department. Debra is an investigator for Adult Protective Services. Both already had retirement in their sights. With a chunk of change in the bank, they have some decisions to make regarding exactly how much longer they want to keep working.
>
> "The freedom is the best thing," said Jack Woodruff.
>
> Before they make any rash decisions, the couple plans to take a cruise and sort some things out.
>
> "We love our life. We have a good life," said Jack Woodruff. "We have been fortunate in a lot of ways. We have a great family." [2]

What exactly does Jack Woodruff mean when he says, "The freedom is the best thing"? It would be a gross misunderstanding to think that Mr. Woodruff meant that he was not free until winning the lottery. There is nothing in the article to suggest he planned to use the money to buy his way out of prison or slavery. Rather, the idea seems to be that the winnings will provide the Woodruff's with *greater* freedom.

Thus, we should note an important distinction: some concepts have both threshold and gradable uses. Consider the concept of "happiness." There is a threshold understanding where we can ask whether someone is happy or not happy. We can also ask whether one person is happier than another, or whether one person was happier at an earlier or later time. The difference between these two understandings can be emphasized by noting that it makes perfect sense to say that Jack is happier than Jill, but neither Jack nor Jill are happy. It also makes perfect sense to say that Jack and Jill are both happy, but Jack is happier than Jill. In the first case, neither is happy enough to make the threshold to be considered "happy," but Jack is closer to the threshold than Jill. In the second case, both meet the threshold standard, but Jack exceeds the standard to a greater degree than Jill.

Applying this to the Woodruffs, we can safely assume that they met the threshold standard for freedom prior to winning the lottery, but the lottery winnings provided them greater freedom. In what sense are they freer? The article suggests an answer: they now have the means to realize some options that were previously beyond them, financially. In particular, it sounds like they will take early retirement to spend more time with their family. So, winning the lottery empowered them to realize options that were not previously available.

It is worthwhile to speculate what might be said if Warren Buffett, rather than the Woodruffs, had won the million dollar lottery. It seems very unlikely that Buffett would claim that the greatest thing about winning the lottery is "freedom." Why the difference? The answer is painfully obvious: Warren Buffett is already worth 60 billion, that is, 60,000 million. After winning the lottery, Buffett would be worth 60,001 million. I am sure Buffett is not averse to gaining another million, but that is about what he makes in 2 hours if he works a 40 hour work week. Buffett has all the money he could spend in dozens of lifetimes, so any effect on his freedom is negligible.

Conversely, suppose the winning ticket provided $10,000 per year for the rest of the Woodruffs' lives. Would Mr. Woodruff still be "over the moon?" No doubt the winnings may have helped, but it is not clear how *much* it would contribute to their overall freedom. Their combined earnings from their regular jobs are likely be in excess of $100,000 per year, so the $10,000 extra would make a comparatively small contribution to their overall financial situation. Compare that now with a single mother earning $8 per hour at a fast-food restaurant. She supports her family on $18,000 a year, $10,000 extra per year would have a significant impact on her freedom: it might allow her to buy a used car or take classes at the local community college, and so on.

Maximizing Freedom: Diminishing Marginal Freedom of Income

We can summarize the previous line of thought by saying that increasing income or wealth provides diminishing marginal freedom. Thus, $10,000 in additional income buys less freedom for the Woodruffs than it does for the single mother earning minimum wage, and $10,000 a year income buys more freedom for the Woodruffs than it does for Warren Buffett.

As with happiness, we can see that redistributing income from the rich to the poor will increase GNF. Taking a billion from Warren Buffett reduces his freedom only slightly. Imagine taking that billion dollars and dividing it into bundles of $10,000 and putting these into paper lunch bags. Suppose further that these bundles are delivered to single parents making minimum wage. On this scenario, 100,000 single-parent families would be the recipients of a lunch bag full of money. As we just noted, this would boost the freedom of the average poor family.

This point bears repeating because it is quite astounding: $10,000 dollars to a poor family increases freedom more than the loss of freedom Warren Buffett would experience if a billion dollars were siphoned off from his bank account. As Lee Iacocca, former chairman of the Chrysler car company, explains:

> Once you reach a certain level in a material way, what more can you do? You can't eat more than three meals a day; you'll kill yourself. You can't wear two suits one over the other. You might now have three cars in your garage—but six! Oh, you can indulge yourself, but only to a point.[3]

Iacocca's rhetorical question "what more can you do?" suggests there is little or no gain in freedom. As he says, yes, you are free to eat more than three meals a day, but this freedom adds nothing of value. As he points out, "you'll kill yourself." Similarly, whether Buffett has 59 or 60 billion is not going to make any difference to how he might spend any day: he can still choose between caviar and champagne or beer and chips, he can buy one car that day or a million. It will not make a difference as to whether he needs to work next year or not. It will not make a difference if he decides to go spend a year on a cruise ship or even buy his very own cruise ship. On the other hand, $10,000 will make a difference for a single parent family: perhaps they might splurge and go to a restaurant, or buy a suit in hopes of improving chances at the next job interview, or buy their first car. The increase in freedom, then, is staggering. Even if we allow that the increase in freedom for one family is the same as the loss of freedom to Warren Buffett, still the fact that 100,000 families would receive a lunch bag with $10,000 shows that the increase in freedom is 100,000 times as much. So, redistributing money from the rich to the poor can have an enormous impact on GNF.

We should consider an objection: if we accept the redistribution argument, then the logical conclusion is to give everyone exactly the same amount of money. To illustrate, imagine the average per-capita income in the United States is approximately $45,000 (which is not too far off from the actual figure).[4] Now imagine any distribution that does not apportion exactly $45,000 to everyone. For example, suppose women are granted $50,000 and men $40,000. This would keep the same average of $45,000, but the resulting income distribution would not maximize total freedom, for our previous argument suggests that every dollar over the average income would buy less freedom for women than redistributing the same income to men. So providing women with $49,000 would reduce their freedom compared with a wage of $50,000, but the resulting loss of freedom would be less than the gain in freedom to men if their wage was boosted from $40,000 to $41,000.

Notice that this is the same objection we considered in the previous chapter: maximum happiness will be achieved when everyone's income is set at the average, that is, at $45,000. But if everyone is paid exactly the same, all incentive to work hard will be gone. With the incentive absent, people will be less productive. With less productivity, less will be produced. And with less produced, there will be a decline in economic output, which will mean the tax base will shrink. In other words, the freedom argument faces the same "failed socialism" objection. Revolutionaries in the twentieth century thought that socialism would make places like the USSR and China better for the masses, but they only made things worse (at least according to this line of objection). Without the profit motive to get people to work, and without the profit motive to allocate effort in an economy, everyone suffers. Despite the rhetoric of the socialists, workers in the United States were much better off than their counterparts in the USSR. They were not better off because they had a chance to become rich capitalists, but simply because the working conditions and material rewards for US workers in the twentieth century were, on average, much better than the conditions in the former Soviet Union.

However, the objection ignores one of the original assumptions of the argument. The assumption was that other things remain equal, and as this objection rightly points out, other things may not remain equal. Indeed, I believe we should accept the point that, in the extreme, there will be reduced incentive to contribute to the economy and, consequently, reduced material output if everyone is paid exactly the same. However, it would be wrong to conclude from this that income should not be distributed differently. What follows is that maximum freedom will be achieved by redistributing some income, but not to the point of perfect equality. The question, then, is how much should the distribution scheme be changed in order to maximize freedom?

Ultimately, the question is empirical: how much monetary incentive is necessary to motivate people to work? To think about how this question might be resolved, consider the tax policy that might go along with the perfect income equality imagined above. Any earning over $45,000 would have to be taxed at 100 percent. At this rate, there is no monetary incentive to work once the $45,000 mark is reached. As we just noted, this would no doubt greatly reduce productivity. The question then is how high the rate might go before we see a decline in productivity. Suppose the tax rate was set at 99 percent for anything above $45,000. Would this provide enough monetary incentive for people to keep working? Suppose you are paid what is presently considered to be a middle class wage of $30 an hour. Once you earn $25,000, an additional pay will net you only 30 cents an hour if taxed at 99 percent. On the other hand, to someone like Warren Buffett, who earns $2 billion a year, a 99 percent tax rate does not look so draconian in absolute terms. Buffett will still earn $20 million a year in this scenario. To put this in perspective, this is the same amount that twenty high school graduates will earn over a *lifetime*.

Economists investigate something analogous to the idea of the previous paragraph under the name, "Laffer Curve." The curve is supposed to explain what the "ideal" tax rate is, where "ideal" is understood as generating the maximum revenue for the government. An important reference source for economists, *The New Palgrave Dictionary of Economics*, reports a wide range of estimates that average out to about a 70 percent tax rate.[5] More recent studies confirm that in the United States, and even in Europe where personal tax rates are much higher, the tax rate could be further raised, garnering more revenue for the government.[6] Since these estimates suggest a tax rate much higher than anything we have considered here, we can safely assume that the financing of BIG through taxes will increase GNF.

Indeed, economists' investigation of the Laffer Curve has several limitations that work in our favor. First, the curve measures two changes in economic behavior: work incentive and tax evasion. Evidence suggests that raising taxes affects tax avoidance much more than it affects work incentive.[7] It is true that some will not work as hard when tax rates go up, but others will work harder in order to keep the lifestyle they are accustomed to. Evidence suggests these two effects cancel each other out.[8] Generally, economists assume that there will be no other change in economic policy than changing the tax rate. Clearly, we might consider increasing penalties for tax cheaters in an effort to slow down some tax evasion and so increase gross tax receipts. A second problem is that the Laffer Curve is too crude a measure for what we are interested in. Generally, rates are compared as national averages. Imagine for the moment that 35 percent is the maximum in terms of gross receipts that the United States can manage. Still, this says nothing about whether tax rates might be reconfigured to increase GNF. Consider that at present the lowest earners pay tax at the rate of 1.5 percent, while the top 1 percent pays about 29 percent.[9] These different rates all contribute to the gross receipts, which is what the Laffer Curve reports on. In terms of increasing gross national freedom, the question would be whether the same 35 percent gross might be achieved by taxing the richest 1 percent at a higher rate, say, 60 percent net, and the lowest income earners at a lower rate, say, –10 percent. As noted, there is overwhelming evidence that the rich could be taxed at a much higher rate with no overall ill effect on the economy.[10] So again, there is ample evidence that redistributing the tax burden to increase the share of the richest class will increase GNF.

Three Concepts of Freedom

We have made a case for redistribution based on an everyday understanding of the term "freedom." In the next few sections, we will compare this sense of the term freedom with competitors. As a preliminary remark, we will, following much precedent, use the terms freedom and liberty (and their cognates) interchangeably.

If there is a single rallying cry among modern political philosophers, from John Locke onward, it is that of freedom and/or liberty. What has been called the "Fundamental Liberal Principle" is the idea that freedom is, or should be, a foundational value in modern societies and that any restriction on freedom, particularly through state coercion, carries a heavy burden of justification.[11] The fact that thinkers as diverse as Locke, Kant, Hegel, Marx, Mill, Rawls, Nozick, and Cohen agree on this principle provides good indication that they likely differ on how liberty is to be understood. For example, Karl Marx thought that communism would increase freedom in the same way that capitalism increased freedom in feudal societies.[12] Nozick, on the other hand, thinks that capitalism provides more freedom than communism. Such radical disagreement suggests that different understandings of freedom are at work. Certainly it would be quite perplexing as to how they could disagree on so much, given their endorsement of the Fundamental Liberal Principle. In any event, in this section we will sketch three different strands of thinking about the nature of freedom. In the following section, we will investigate how these compare with our everyday understanding of freedom.

NegativeL iberty

What has become known as "negative freedom" or "negative liberty" is the concept of noninterference by others, that is, noninterference by individuals, groups, or the state. Hobbes provides a classic definition of this understanding of liberty:

> By "liberty" is understood, according to the proper signification of the word, the absence of external impediments; which impediments may oft take away part of a man's power to do what he would, but cannot hinder him from using the power left him according as his judgment and reason shall dictate to him.[13]

Negative liberty so understood makes sense of a number of everyday examples where we would say there is a loss of freedom. A law that prohibits me from smoking marijuana is a loss of freedom in the negative sense of liberty. If I get caught smoking pot, I could face jail time, and certainly being in jail counts as an external impediment. Similarly, the requirement to pay taxes looks to be an external impediment in Hobbes's sense: if I do not pay my tax bill, I could end up in jail. The negative sense of freedom also makes sense of non-state actors limiting my freedom: if you kidnap me, there is an external impediment to my freedom. If you park your car across my driveway so I cannot drive my car, there is an external impediment to my freedom.

Negative freedom is sometimes referred to as "formal freedom" because proponents of negative freedom typically see the best way to enshrine negative freedom is through laws that limit interference. A legal, or perhaps, even better, a constitutional right to freedom of expression is conceived in the negative freedom tradition as a legal protection against interference with expression. A legal, or perhaps a constitutional, right to freedom of trade is conceived in the negative freedom tradition as a legal protection against interference with the buying and selling of property between contracting parties. Similar points would apply to legal guarantees of noninterference in religious matters and peaceful assembly.

Proponents of negative liberty as a foundational value in political philosophy seem to have an obvious and compelling response to the diminishing marginal freedom of income argument. The Woodruffs, the single parent, and Warren Buffett, all face the same external impediments to liberty, hence they are equally free. Indeed, according to this line of thought, a homeless person and Warren Buffett have the same amount of negative freedom. A slave or a convicted felon has more external impediments, for sure, but in our example, no such external impediments are present. So, in the Woodruff's claim, they are using "freedom" in a different sense than "negative liberty." After all, the legal or constitutional rights of the Woodruffs did not change after winning the lottery: they enjoyed the same right to freedom of expression and freedom of trade before and after winning the lottery, hence their negative freedom remains unchanged.[14]

Positive Liberty As Authentic Desire

Unsurprisingly, negative liberty is often contrasted with positive liberty. Sir Isaiah Berlin famously made the distinction thus:

> The first of these political senses of freedom or liberty (I shall use both words to mean the same), which (following much precedent) I shall call the "negative" sense, is involved in the answer to the question "What is the area within which the subject—a person or group of persons—is or should be left to do or be what he is able to do or be, without interference by other persons?" The second, which I shall call the "positive" sense, is involved in the answer to the question "What, or who, is the source of control or interference that can determine someone to do, or be, this rather than that?" The two questions are clearly different, even though the answers to them may overlap.[15]

On Berlin's distinction, the scope of the negative freedom of the single parent, of the Woodruffs, and of Buffett is the same, since they are all subject to the same laws governing when others can and cannot interfere with them. Positive freedom, as Berlin sees it, has to do with the "source" of control or interference in our actions. Thomas Hill Green is emblematic of the positive conception of freedom that emphasizes self-control of our actions; for example, someone who washes his hands a hundred times a day may be free from external impediments, but there is a sense in which this person is not free according to the self-control conception of freedom.[16] Similarly, we might say a drug addict is not free, even in jurisdictions where there are no laws against drugs. Green compares the drug addict to a slave.[17] The internal compulsion to take drugs restricts the drug addict's freedom in much the same way that one's freedom would be reduced if one were forced to take drugs.

Positive Liberty: The Power to Act

A second sense of positive freedom emphasizes having the power to act. There are no laws, as far as I know, prohibiting me from jumping over the moon. Similarly, there are no laws against inviting the entire student body of my university on a weeklong cruise at my expense. According to the positive conception of freedom as the power to act, there is a very real sense that I am not free to jump over the moon, nor am I free to take the 20,000 students at my university on a cruise. I lack the power to jump over the moon, so even though there are no legal restrictions, I am not free to jump over the moon. The reason I am not free to take the student body on a cruise is that such a cruise would cost about 20 million dollars and I am a little over 20 million short of having sufficient funds. The positive conception of freedom as power to act says that, in this case, I am not free to take the students on a cruise since I lack the financial resources. I lack the power to act in this way, and so I am not free in this respect.

For our purposes, the most relevant point is that limits to economic power mean limits to freedom. Tawney points out that failing to recognize the economic dimension to freedom means that the freedom of some is purchased at the price of the freedom of others. As Tawney explains:

> If liberty means, therefore, that every individual shall be free, according to his opportunities, to indulge without limit his appetite for either, it is clearly incompatible, not only with economic and social, but with civil and political, equality, which also prevent the strong exploiting to the full the advantages of their strength, and, indeed, with any habit of life save that of the Cyclops. But freedom for the pike is death for the minnows.[18]

Against mere negative liberty, Tawney writes that liberty must be construed "realistically" as including economic freedom:

> Hence, when liberty is construed, realistically, as implying, not merely a minimum of civil and political rights, but securities that the economically weak will not be at the mercy of the economically strong, and that the control of those aspects of economic life by which all are affected will be amenable, in the last resort, to the will of all, a large measure of equality, so far from being inimical to liberty, is essential to it. In conditions which impose co-operative, rather than merely individual, effort, liberty is, in fact, equality in action, in the sense, not that all men perform identical functions or wield the same degree

of power, but that all men are equally protected against the abuse of power, and equally entitled to insist that power shall be used, not for personal ends, but for the general advantage.[19]

Parable: The Hole Truth About Freedom

Which of the three conceptions of freedom should we adopt for the purposes of political philosophy? We will spend some time investigating this question. In this section, I hope to show that at least in their paradigmatic formulations, each differs significantly from our everyday understanding of "freedom." We will start with a little parable:

The Hole Truth About Freedom:

Hobbes, Green and Tawney are out for a walk in the woods. Their conversation starts with some personal matters, e.g., they find much to their surprise, that all three have a phobia of using ladders. They marvel and laugh at the coincidence. Soon the conversation turns to more serious matters. When discussing the notion of freedom, the three quickly realized that they have radically different understandings. Hobbes endorses the negative notion, Green, the autonomy sense, and Tawney believes that freedom means effective power to act. To illustrate his view, Hobbes points out that laws restrain freedom. As an example, he cites a local ordinance that prohibits people from climbing out of holes and assisting others from climbing out of holes. The law, apparently, was put in place several hundred years ago, motivated by religious beliefs. It was thought that falling in a hole was a sign from God, and that to try to escape was to work against the will of God.

Naturally, as they are in heated discussion, they become oblivious to their surroundings and the three of them fall into separate holes. Each shouts for help, but they quickly realize that there is no one there to rescue them. As luck would have it, however, a philosophical god heard their conversation about the nature of freedom. The philosophy god promises to free them. An hour later the philosophy god returns with some parchment, Sigmund Freud, and a ladder. On the parchment is written an order from the local council suspending the law against climbing out of holes for Hobbes, and Hobbes alone. The god lowers the parchment to Hobbes. From the edge of the hole, Sigmund Freud is able to talk to Green and help Green overcome his phobia of climbing ladders. Finally, the god lowers the ladder down into the hole where Tawney is trapped.

By this point, a large crowd from the town has gathered. The god announces, to much applause, that he set each of the theorists free according to their preferred conception of freedom. The village idiot then spoke up, "I don't understand. I don't see how they are free.

> Yes, Hobbes had the external restraint of the law removed, but he still lacks the power to act, necessary to get out of the hole, and he still has an internal constraint: his phobia of using ladders. Yes, Green has had his phobia removed by Dr. Freud, but he still lacks the material means necessary to get out of the hole and he still faces prosecution if he climbs out of the hole. Finally, Tawney does not seem free because he too will face prosecution if he climbs out using the ladder, and he too has an internal obstacle to his freedom, namely, his phobia." Unsurprisingly, the philosophical sophisticates in the crowd laugh at the village idiot's naiveté.

For our purposes, the only significant point I want to draw from the parable of the hole truth is that our everyday conception of freedom is not identical to the three political conceptions of freedom outlined above. Although not identical, each seems to capture something important about our everyday notion. Proponents of negative liberty are right that arbitrary laws, like prohibitions on climbing out of holes, constrain our freedom. To say "you are free to do X, but as soon as you do X, you will go to jail" is not an argument that you are free to do X, but an argument that you are not free to do X. So, giving Tawney and Green a ladder and therapy to overcome their phobias of climbing ladders is not enough for freedom, they require the external constraint of the law to be removed as well before they can be said to be free. Green is right that internal constraints also count against freedom. So, granting Hobbes and Tawney clemency from prosecution and a ladder are not enough for their freedom, they need the phobia removed before they can climb to freedom. Finally, Tawney is right that freedom is intimately related to power. So, granting Hobbes and Green clemency from prosecution and therapy to overcome their phobias is not enough, they require an effective means to achieve freedom, namely, a ladder.

I should perhaps add that the parable is not offered as a serious exposition of the views of Hobbes, Green, and Tawney on freedom. Indeed, as an exposition of their views, the parable is simply false. Green and Tawney explicitly allow that the negative conception of liberty holds an important truth. What Green and Tawney object to is the idea that negative freedom is a sufficient exposition of freedom.

The Plea for Negative Liberty

It should be apparent that it is the "ability to act" conception of freedom that the Woodruff's appeal to when they say of the lottery, "the freedom is the best thing." The lottery winnings provide them with power to act in ways that they lacked formerly, they can retire early and travel, for instance. But this raises an important concern, perhaps the everyday conception of freedom is not the one we should use for political philosophy. Rather, in political philosophy, we should understand the term "freedom" as "negative freedom." The objection is important, because if it is successful, it would undermine our earlier argument. The Woodruffs did not gain freedom when they won the lottery if "freedom" is understood in the negative sense.[20] They were still subject to the same external constraints as before. They did not gain, for example, the freedom of religion or the freedom of speech, they had these freedoms before their winnings. So, in the political sense, they are as free (or unfree) as they were before winning the lottery.

The following passage from Friedrich Hayek defends the idea that negative freedom is the important concept for political philosophy:

> In the United States it has come to be widely accepted as the foundation for the political philosophy dominant in "liberal" circles. Such recognized intellectual leaders of the "progressives" as J. R. Commons and John Dewey have spread an ideology in which "liberty is power, effective power to do specific things" and the "demand of liberty is the demand for power," while the absence of coercion is merely "the negative side of freedom" and "is to be prized only as a means to Freedom which is power."
>
> This confusion of liberty as power with liberty in its original meaning inevitably leads to the identification of liberty with wealth; and this makes it possible to exploit all the appeal which the word "liberty" carries in the support for a demand for the redistribution of wealth. Yet, though freedom and wealth are both good things which most of us desire and though we often need both to obtain what we wish, they still remain different.[21]

Hayek's charge, then, is that "progressives" confuse two senses of freedom: the negative conception and the sense of freedom as power. As Hayek himself notes, this is disputed by his opponents who understand the negative conception to be merely a special case of the idea of liberty as the power to act. Nevertheless, let us stipulate that Hayek is correct that there really are two different concepts at work here. The obvious reply to Hayek's objection, then, is that if there really are two conceptions, proponents of liberty should promote and defend both. Hayek's defense against this seems to be simply that "liberty" originally meant negative liberty. But so what? If that is what it meant originally, then the original meaning is too narrow, and we should endorse and defend both liberty in its original meaning and in the meaning of "freedom as power."

An analogy might help. Libby says, "We should not go to war, we should love our brothers." Fred replies, "There is a confusion here between the meaning of 'brother' as 'male sibling' and 'brother' as meaning 'all members of the human race.' The original meaning is 'male sibling.'" Libby replies, "Well, I believe that the former definition is encompassed by the latter definition. However, if these are different concepts, then I endorse both: you should love your male siblings and all of humanity." Of course Libby's point is analogous to Hayek's progressive opponent's point, and Fred's point is analogous to Hayek's point. It should be clear that the argument is not advanced by merely noting there are distinct concepts at work. To defend the primacy of negative liberty, some argument has to be made that it is only the negative freedom concept that we should hold in political philosophy.

A more promising line of objection is offered by Berlin:

> It is important to discriminate between liberty and the conditions of its exercise. If a man is too poor or too ignorant or too feeble to make use of his *legal rights*, the liberty that these rights confer upon him is nothing to him, but it is not thereby annihilated. The obligation to promote education, health, justice, to raise standards of living, to provide opportunity for the growth of the arts and the sciences, to prevent reactionary political or social or legal policies or arbitrary inequalities, is not made less stringent because it is not necessarily directed to the promotion of *liberty* itself, but to conditions in which alone its possession is of value, or to values which may be independent of it.[22]

Suppose then we compare the freedom of a homeless person in the United States to Warren Buffett and ask who is freer. The answer, according to Berlin's understanding of freedom, is that they are both equally free. John Rawls offers a similar line of thinking:

> Many have argued, particularly radical democrats and socialists, that while it may appear that citizens are effectively equal, the social and economic inequalities likely to arise if the basic structure includes the basic liberties and fair equality of opportunity are too large. Those with greater responsibility and wealth can control the course of legislation to their advantage. To answer this question, let us distinguish between the basic liberties and the worth of these liberties as follows: the basic liberties are specified by institutional rights and duties that

> entitle citizens to do various things, if they wish, and that forbid others to interfere. The basic liberties are a framework of legally protected paths and opportunities. Of course, ignorance and poverty, and the lack of material means generally, prevent people from exercising their right and from taking advantage of these openings. But rather than counting these and similar obstacles as restricting a person's liberty, we count them as affecting the worth of liberty, that is, the usefulness to persons of their liberties.[23]

So, using our previous example, Berlin and Rawls will say that the homeless person and Buffett have equal freedom, but the worth or value of freedom is much greater for Buffett than the homeless person. The homeless person can do much less with his freedom, and so it is worth much less to him than Buffett's freedom is worth to Buffett.

Parable: Three Islands

The idea that we ought to distinguish freedom and the value of freedom, as Berlin and Rawls suggest, I shall argue, is not very satisfying. We saw in the parable of the Hole Truth About Freedom that there is a real disconnect with our everyday understanding of liberty and freedom. On Rawls's view, for example, we would have to say that Jack, who is stuck in a hole because he cannot afford to pay to get out, is as free as Jill, who is not stuck in a hole. Rather, the difference is simply the value of Jack's and Jill's freedom. True, Rawls would not be indifferent to the plight of Jack. Rawls might say that Jack should have enough income, a social primary good, in order to escape the hole. And so Jack's plight is one of injustice. Suppose Rawls is correct about the distributive injustice that Jack faces, still, it is utterly bizarre to say that Jack has just as much freedom as Jill. This in itself should make us think that Rawls has significantly redefined a key political term.

The following parable of the Three Islands makes the same point about redefining freedom, but this time on a societal level.

> Your boat sank. Fortunately you were able to get into your dingy. Not far are three islands surrounded by coral reefs. Your only hope of getting to land safely is to row your boat to the edge of one of the reefs and then get out and swim to the beach. This means that once you decide which island to land on, your choice will be irrevocable, since your dingy will be washed away by the tide. Here are the possible landing spots:

A-Island. A-Island guarantees freedom of religion and freedom of assembly. Legislation generally favors freedom of expression with the exception that racist comments and other forms of hate speech are not permitted in public fora, nor in private meetings with more than three persons present. One is permitted to make racist comments in private, and write racist literature for private consumption. Legislation guarantees a small plot of land to all 110 inhabitants. Each plot is beachfront, with sufficient coconuts and mollusks to keep a person alive. Through legislative restriction, the plots can neither be bought nor sold. Some of the inhabitants of A-Island have built elaborate shelters on their plots, some spend their days writing novels, others playing musical instruments, some train and play sports, some are content to live simply and lie on the beach most days, and others enjoy surfing.

B-Island. B-Island is similar to A-Island in that B-Island guarantees freedom of religion and freedom of assembly. Legislation generally favors freedom of expression with the exception that racist comments and other forms of hate speech are not permitted in public fora, but they are permitted in private meetings with no restriction on the number in attendance. One is also permitted to write racist literature for private consumption. B-Island has a freer economy in that there are no restrictions on buying and selling land. When the residents of B-Island first came to B-island, the land was divided equally among all, just like it was on A-Island. Over time, however, the preferred land ended up in the hands of just 10. Some lost their land in card games, others sold their land for alcohol, and others traded their land for sexual favors. None of the exchanges of land for goods or services was coerced or involved fraud. Of the 100 who do not own any of the preferred land, 50 work for the land owners. They perform such tasks as collecting and opening coconuts, building and maintaining shelters, collecting freshwater, and performing sexual services (agreed by contract). Most of the workers work 14 hours a day. A few of the more talented have negotiated more favorable contracts, allowing them to work only a few hours a day. In return, they are paid in coconuts and mollusks and are provided a place to sleep. The 50 who do not have land and who do not work for the land owners eke out an existence on tidal flats and a large rock. At high tide, all 50 must cram together on the highest part of the rock in order to avoid being swept to sea. At low tide, they collect low digestibility seaweed that keeps them just above the point of starvation. Most are kept going by the hope that they might get a job on the main part of the island.

C-Island: C-Island is the same as B-island in terms of the ownership of land by 10 individuals, 50 workers, and 50 eking out their existence on a giant rock at high tide. The one difference is that there is no restriction on freedom of expression: hate speech is permitted in public fora.

Your trouble, as you sit in your dingy, is that the islands are indistinguishable from a distance and you have lost your map. You know the different socioeconomic structure of each island, but you do not know which of the three islands is B-Island, where you hope to retire. You reason that this island will give you the most freedom. You recently had property left to you by relatives on B-Island and C-Island. However, you lost your land on C-Island in a card game and so you want to avoid C-Island at all costs. B-Island will allow you to order around five servants as instruments of your wants and pleasures. Although much rides on your choice, you realize that the best you can do is to choose an island at random and hope for the best. As you begin to row towards one island chosen at random, the ghost of John Rawls appears and says, "You are rowing toward the wrong island, at least if you care about freedom. You should head to the island on your far left if you value your freedom and the freedom of society at large." You ask, "Do you mean the one on the far left is B-island?" However, the ghost vanishes as you ask your question. While it is a shame that the eminent philosopher did not answer your question, still, you think it is best to take his advice rather than letting luck be your guide. After all, John Rawls was known as a kind and generous person.

When you hit the shores of the far left island, it soon becomes apparent that you have landed on C-Island. During low tide you scrounge seaweed and whatever else you can find to eat in the tidal flats. At high tide, you must crush next to 49 others at the top of the rock as you attempt to avoid being swept into the ocean. You dream of one day gaining employment and enjoying the relative comfort of working 14 hours a day, seven days a week for one of the land owners. After two months of this, you decide to exit this world. As you are about to throw yourself from the rock to the jagged coral reef below you cry out, "Damn you ghost of John Rawls. You tricked me. This is not the island of freedom. I have no choice about how my life goes if I want to live. Twice a day I must clutch and huddle on a rock and twice a day I must scrounge as best I can to eke out an existence. This is not freedom; this is worse than slavery." The ghost reappears and says, "You are mistaken my friend. A-Island has less freedom because legislation forbids the buying and selling of land and forbids certain forms of hate speech. B-Island has less freedom because the laws forbid racist speech in public fora. Certainly you would have been much happier on A-Island or B-Island, but that is not what we were talking about. Nor are we talking about justice. Clearly, the land owning democracy of A-Island most closely resembles the ideal I put forward in A Theory of Justice, but again, this is not what we were talking about. We were talking about freedom."

From the point of view of the everyday understanding of freedom, the ghost of John Rawls's conception of freedom is a perversion of what is meant by freedom. It is incomprehensible how C-Island offers you the most freedom. If the reply by Rawls is that it is a stipulative definition of freedom, then it is an almost Orwellian redefinition, where "unfreedom" means freedom. It is bizarre to think that this can be papered over by the thought that there is more freedom on C-Island, and that the freedom on C-Island is worth more to some and less to others. It is not the value of freedom that has changed, *it is freedom itself that is diminished.* Note that I am not saying Rawls would endorse C-Island as the one that best approximates the good society. As intimated, Rawls would probably say that A-Island most closely approximates what he envisioned in *A Theory of Justice*, at least more so than B-Island or C-Island. The point is that the way Rawls understands freedom does not allow him to make the judgment that we intuitively want to make, that is, at least one thing wrong with B-Island and C-Island, in addition to whatever else might be wrong, is that 10 people on both B-Island and C-Island have a lot more freedom than the other 100 inhabitants. Indeed, remember that part of the reason for wanting to go to B-Island in the first place was because of the freedom it offered to live your life as you pleased. With five servants at your beck and call, you have instruments to assist you in fashioning your life in any number of directions.

Rawls's understanding of freedom is doubly mistaken, as it gives the wrong rankings of the overall freedom of each island, and also the wrong ranking of how much freedom each island offers you as an individual. Rawls must give the following rankings of the islands in terms of overall societal freedom: C-Island is the freest, B-Island is second, and A-Island is the least free. A has more restrictions on freedom of expression than either B or C. He must give the same rankings for you as an individual: you face more legal restrictions on freedom of expression on A-Island and the fewest on C-Island, so you have more freedom on C-Island. True, Rawls may say that your freedom is worth more to you on B-Island or C-Island, but this is not what we are asking. On an everyday understanding of freedom, where we mean being able to determine the direction of your life, A-Island has the most freedom overall. True, the citizens there are restricted from criticizing one another in certain hateful ways, and there are limits on trading resources, but, on average, the citizens of A-Island have the most scope to determine the direction of their lives. If ranking these islands in terms of your personal freedom, then B-Island offers you the most freedom, followed by A-Island and finally C-Island. The fact that Rawls's understanding of freedom gives such counterintuitive rankings suggests that something has gone terribly wrong.

The Law and Freedom

It may be protested that something must have gone wrong with the argument, since it seems to suggest that more governmental restrictions equals more freedom. Consider that on A-Island, there are restrictions on free trade that are absent on B-Island and C-Island. This seems to violate, in a very serious way, the catechism among certain political groups in the United States (and other nations), captured by the bumper sticker "Less Government = More Freedom." We can see how this line of thought is naturally suggested by the negative conception of freedom:

> Laws restrain behavior. In the typical case they are like a promise to interfere if certain actions are taken. So, laws restrain freedom. The less government restraint by law, the more freedom.

I say "in the typical case" because laws that restrain other laws are a possible exception: a law or right to free speech is a restraint or promise to *not* make laws limiting expression. However, most laws are much more like a promise to interfere: if you steal, we will interfere with you and put you in jail. If you do not pay your taxes, we will put you in jail.

Legal restrictions, then, seem like constraints on freedom, and lifting legal red tape seems like a way to answer the call of freedom. To see this thinking in action, consider this quote from *The Washington Post*, "At campaign stops, Sullivan repeats this catchy phrase, 'More freedom, less government.'"[24] The phrase is certainly catchy.

Yet, there is a problem when we take this thinking to its logical conclusion: maximal freedom is where there is a complete absence of law. In the minds of many who spout the "less government = more freedom" rhetoric, such a situation is not one where there is more freedom but complete anarchy. It invites images of postapocalyptic wastelands terrorized by roving bands of savage bikers wearing skull masks. A society absent of government agents, like the police and an army, does not sound like one of more freedom but of less. There is little freedom when one is constantly fearful of robbery, violence, and death. Surely, it is not always the case that less government means more freedom.

John Locke takes the opposing view—laws aim to expand freedom:

> So that, however it may be mistaken, the end of law is not to abolish or restrain, but to preserve and enlarge freedom: for in all the states of created beings capable of laws, where there is no law, there is no freedom: for liberty is, to be free from restraint and violence from others; which cannot be, where there is no law: but freedom is not, as we are told, a liberty for every man to do what he lists: (for who could be free, when every other man's humour might domineer over him?) but a liberty to dispose, and order as he lists, his person, actions, possessions, and his whole property, within the allowance of those laws under which he is, and therein not to be subject to the arbitrary will of another, but freely follow his own.[25]

What should we make of Locke's suggestion that laws may actually "enlarge freedom" rather than diminish it? Surely it seems to be very counterintuitive to think that laws do not diminish freedom, even if we never wanted to use the freedom in question. Perhaps you will go through your life never wanting to murder someone else, but surely it seems that laws diminish your freedom when it comes to murder. Locke's answer seems to be that without certain laws, like the law against murder, we would be less free because others might interfere with us, that is, otherwise we would be "subject to the arbitrary will of another." But this is consistent with saying that laws do diminish freedom in certain respects, for example, the freedom to murder; but it increase freedom overall: with laws against murder, I feel confident to leave my home rather than stay behind barricaded doors.

The fact that laws constrain my freedom, even if they contribute to overall freedom, is evident from the fact that Locke offers us a false dilemma. Imagine one were king of the United States such that one could make laws that apply to everyone but the king. Here, one would have maximal amount of freedom: no laws binding one's own will, but also freedom against the "arbitrary will of another." As king, immune from the law, one would have the freedom to murder, and set up factories spewing toxic waste next to orphanages, and so on. Here the slogan for the king is, "less government for me, and more for you." Most laws have this dual aspect to them, they diminish freedom in one respect and enlarge it in another. The example of the king makes this point clear, since the king is not subject to the laws, he is free from the restraining part of laws, whereas he enjoys the full benefits of the freedom-enhancing aspects of laws. Of course, for the rest of us, we must live with this dual aspect of laws, that laws both diminish and enhance freedom.

The "less government = more freedom" rhetoric gains plausibility only so long as we forget this dual aspect of laws. Indeed, the rhetoric may seem particularly appealing because it is much easier to see when laws restrict our freedom than when they enlarge our freedom. Suppose your neighbor calls the police when your party violates bylaws governing loud music past 10 o'clock. Annoyed at this restriction on your freedom, you successfully petition local council to repeal the noise laws. Now your freedom is enlarged—at least, so you think. However, your neighbor too has his freedom enlarged. Now he can cut his lawn at 5 a.m. every weekend morning and he takes full advantage of this fact, using the nosiest lawnmower money can buy. On Saturdays, he mows it in a north to south direction. On Sundays, he mows it in an east to west direction to ensure lawn manicure excellence. You have now lost the freedom to sleep in on the weekend. On reflection, you decide your freedom has diminished overall. Yes, your freedom to have loud parties is enhanced, and you may take advantage of this several times a year, but this gain in freedom is more than outweighed by the loss of freedom to sleep-in every single weekend. Those who think that less government automatically means more freedom ought to be careful what they wish for.

So we can happily admit that the restriction on selling one's plot of land on A-Island does decrease freedom along one dimension, just as laws against murder restrict our freedom. However, the restriction results in more freedom in total on A-Island, as opposed to B-Island and C-Island. The connection with the main argument of this work is perhaps obvious. Introducing a VAT tax or reconfiguring the tax code to shut tax loopholes decreases our freedom with respect to paying taxes, but it will increase the freedom of society as a whole. The fact that many laws have this dual aspect—decreasing freedom along one dimension, but increasing freedom along another—shows that the slogan "less government, more freedom" does not bear reflective scrutiny: it is simply false when understood as a universal claim.

Just to be clear, the claim here is not that every law necessarily increases overall freedom. A king could impose on his subjects draconian taxation, legal restrictions on speech, assembly, and so on in such a way as to greatly reduce the freedom of society as a whole. The point, rather, is that some (perhaps many) laws increase overall freedom.

Freedom as Substantive Autonomy: The Rallying Cry of Modernity

The preceding discussion of the three conceptions of freedom suggests the following result: freedom as autonomy and freedom as effective power are the central notions. Negative liberty is an important means to achieve autonomy and effect power. To support the contention further, it will be helpful to consider the nature of autonomy first.

John Christman nicely summarizes much current work on autonomy: "Put most simply, to be autonomous is to be one's own person, to be directed by considerations, desires, conditions, and characteristics that are not simply imposed externally upon one, but are part of what can somehow be considered one's authentic self."[26] Much of the voluminous theoretical literature on autonomy has concentrated on what has become known as "competency"' and "authenticity"; the conditions for autonomy. Competency conditions are said to include such things as "rational thought, self-control, and freedom from debilitating pathologies, systematic self-deception, and so on."[27] Authenticity conditions are often explicated in terms of the ability to reflect upon and endorse one's values and desires. The best-known account of authenticity comes from Harry Frankfurt, who suggests that autonomy requires that our second order desires (our desires about our desires) are such that the first order desires should issue into action.[28] Drug addicts who wish to kick the habit are paradigm examples of cases where second order desires are not aligned with first order desires. Such addicts desire drugs, but desire that they do not have the desire for drugs.

The focus on competency and authenticity conditions is important, but these need to be supplemented with some concern about "external conditions" for autonomy. This point may be appreciated by reflecting on the analogy made between political autonomy and individual autonomy. A state is said to be autonomous when it is able to rule itself rather than be ruled by a foreign power. Thus, for example, the American Revolution was a cry for autonomy: the colonies rejected the rule of Great Britain. To the extent that Great Britain influenced political power in the colonies, the colonies were not autonomous. The colonies lacked the ability to govern themselves. Indeed, the very etymological root of "autonomy" means "self-lawgiver." But then consider the plight of a slave. The slave's life is determined by the commands of another. The slave does not shape the course and direction of his life, the master does. Notice, however, that a slave may meet the competency condition of autonomy. The slave is rational, exhibits self-control, is free from debilitating pathologies and self-deception, and meets the authenticity condition; for example, the slave is able to reflect and endorse his desires (including, presumably, the desire to be free). If one held that autonomy only involves the internal conditions, then one would have to admit that such a slave was autonomous.

In any event, in order to avoid dispute about whether autonomy does in fact require something more than internal conditions as some authors have suggested,[29] we will simply stipulate that "substantive autonomy" requires that some external conditions be met. Among these conditions are that the agents are not enslaved or incarcerated. Agents are more substantively autonomous to the extent that they have the power to realize a range of life options. That is, other things being equal, the more options one has within one's power to act on, the more one is autonomous. Thus, there are a number of ways in which the external conditions of substantive autonomy might be compromised. If you lose your limbs to flesh eating bacteria, you are less substantively autonomous (other things being equal). If you lose one or more of your senses, then you are less substantively autonomous (other things being equal). As the parable of the hole truth illustrates, if you are stuck down a hole, your substantive autonomy is reduced. And as the parable of the three islands indicates, if you are economically powerless, then you have reduced autonomy compared to the economically empowered (other things being equal).

It may help to think about the relationship between substantive autonomy and negative liberty in historical terms. Part of the protest of the Enlightenment was against state and church oppression of the rising middle class. The call for freedom of expression, for example, was a reaction against heavy state censorship. The call for freedom of religion was a reaction against laws prohibiting any but the state mandated religion or religions. Calls for freedom of trade were a response to feudal restrictions tying workers to the land, and monopolies being granted by the king or queen to the favored few. However, we should understand the call for these negative freedoms as a means to secure substantive autonomy for society at large, not just the favored few. The aspiration of the Enlightenment is that we should be authors of our own lives, not mere puppets in a drama ordained by church and state.

Certainly these negative freedoms, freedom from interference of the state into our religious, cognitive, and economic lives, offered more substantive autonomy to the general population as compared with feudalism. The fact that one was not born tied to some particular land and lord, to a particular religion, or to state-approved reading materials, was a boon for substantive autonomy. Notice, however, the structure of this line of thought: negative liberty is used as an *instrument* to promote or preserve substantive autonomy.

If it is correct that negative freedoms are tools for promoting substantive autonomy, then these tools can be evaluated in terms of how effective they are in achieving their ends. Laws permitting freedom of expression have, by and large, been successful in promoting substantive autonomy. The market place of ideas, as Mill termed it, has helped individuals explore a variety of viewpoints. It need not have turned out this way. It is possible to imagine Dr. Mesmer with the power of hypnotic speech such that he can hypnotize anyone into doing anything. As millions under the hypnotic influence line up to give away all their possessions to the doctor, we might insist that he refrain from using his hypnotic speech. If he claims that this goes against our love of freedom, we should respond that it is our love of freedom that requires us to limit his speech. His speech undermines the substantive autonomy of millions. The point, then, is that there is no guarantee that negative liberty, as a tool for achieving substantive autonomy, will achieve the desired aim, since all tools may fail to bring about their intended results.

I suspect the idea that a system of negative liberty will automatically ensure that the resulting society will be free is due to a seductive but mistaken line of reasoning.

1. Suppose all individuals at time T1 in some society are free (substantively autonomous) and granted negative rights, including the right to buy and sell in an open market with each other at agreed upon prices.
2. At times T2 to T4, individuals buy and sell goods and labor in a noncoerced fashion.
3. At time T5, the individuals in this society must be as free (substantively autonomous) as they were at T1.

The argument has a certain intuitive appeal. If people are free to begin with, as 1 states, and no one is forced to buy or sell anything, as 2 states, it seems hard to imagine how people could end up less free. Where, it might be asked, does non-freedom enter if not at either step

1 or step 2? Since, by hypothesis, there is freedom in both steps, there must be as much freedom at the final state described by 3.

The problem, as we have seen repeatedly, is with step 2. At best, 2 guarantees that there are free exchanges, not that these exchanges are freedom (substantive autonomy) preserving. In other words, 2 guarantees a free *process*; 2 does not guarantee freedom as a *product*. Recall in chapter 3 we discussed the example of the husband who agrees to become the personal slave to a surgeon in exchange for a lifesaving operation for his wife and the mother of his children. If the family is financially destitute and has no other means to save the woman, such an exchange seems perfectly rational. The prohibition on slavery is clear acknowledgement that sometimes free exchanges can lead to less freedom. When we prohibit the husband from saving his wife in this way, we force him to be free.[30]

We noted too in the example of the waterhole that it could happen that, through a number of free exchanges, one person could end up having a monopoly on a vital resource like water, and enforce slave-like conditions on the rest of the population. We noted too that without positive and negative legislation, monopolies would tend to develop. In such cases, in order to preserve freedom (substantive autonomy), we have to limit certain free exchanges. This is illustrated with the case of A-Island where the negative right to sell land is limited but yet A-Island enjoys more freedom (substantive autonomy).[31]

The idea of substantive autonomy explains the Woodruff's claim that winning the lottery increases their freedom. They have the power to follow a much wider variety of life plans with their new wealth. Indeed, the idea of substantive autonomy seems pretty close to what nonphilosophers mean by "freedom," for on this assumption, we can explain why, in the parable of the hole truth of freedom and the parable of the three islands, freedom is absent.

Measuring Freedom

The claim that money tends to increase freedom, conceived of as substantive autonomy, depends on the assumption that we have some way of measuring freedom. For example, the Woodruff's claim that they have more freedom now as a result of winning the lottery presupposes some means of comparing the amount of freedom before winning the lottery and the amount of freedom after winning the lottery. A moment's reflection reveals that this is no easy task.

Indeed, according to Ian Carter, there is a division among liberal thinkers about the notion of overall freedom:

> For some libertarian and liberal egalitarian theorists, freedom is valuable as such. This suggests that more freedom is better than less (at least *ceteris paribus*), and that freedom is one of those goods that a liberal society ought to distribute in a certain way among individuals. For other liberal theorists, like Ronald Dworkin (1977, 2011) and the later Rawls (1991), freedom is not valuable as such, and all claims about maximal or equal freedom ought to be interpreted not as literal references to a quantitative good called "liberty" but as elliptical references to the adequacy of lists of certain particular liberties, or types of liberties, selected on the basis of values other than liberty itself. Generally speaking, only the first group of theorists finds the notion of overall freedom interesting.[32]

We side with the former group who find the notion of overall freedom interesting. Skepticism about overall freedom is sometimes motivated by examples like the following. Suppose half a million citizens live on the island nation Newtopia. They enjoy a BIG of $10,000 a year from an endowment provided by the rich founder of Newtopia. The endowment comes with a stipulation that the law regulating speech against negative statements made about Martians is upheld. The law states that no one should speak ill of Martians. Penalties include fines and imprisonment for repeat offenders. Do the residents of Newtopia have more or less substantive autonomy than citizens in the United States? One thing that makes the question particularly difficult to answer is that it asks us to weigh freedom along two different vectors: the freedom BIG provides versus a limitation in freedom of speech. In this case, I am inclined to think that citizens of Newtopia have more freedom than their American counterparts because the loss of freedom of speech is very minimal, whereas the gain in economic freedom is quite substantial.

However, to say that we can measure freedom does not mean that we can measure it in all circumstances. It may, for example, be impossible to decide in certain cases where trade-offs must be made. If the endowment was for merely one dollar per citizen per decade, then such a restriction on freedom of expression would clearly be a loss of overall freedom. As we raise the financial stakes, we may not know where the loss of freedom of expression might be compensated for by the increased financial freedom. However, to show that we cannot always determine some value for overall freedom does not mean that it is never measurable, or that there is no such thing as overall freedom.

Two further reasons support the idea that the claim that BIG will increase GNF requires fairly modest assumptions about measuring freedom.

First, the discussion of BIG is a single vector question: does redistributing income along the lines suggested by BIG increase GNF? Here we need only compare the effect on substantive autonomy of taxing the rich with its effect on the substantive autonomy of giving to the poor. It is a single vector, then, because it asks only about the relationship between income and freedom. In answering this, we assume the other vectors of substantive freedom remain unchanged. In general, two vector questions, like the question of the freedom of the citizens of Newtopia, appear to be much harder to answer.

A second point is that there is no commitment here to the idea that measurements of freedom must be cardinal. Cardinal measures have a consistent meaning or unit between any two intervals. So, for example, it makes sense to say that someone three feet tall is half as tall as someone six feet tall, and that the height difference between a four foot tall and a five foot tall person is the same as the difference between a three foot tall person and a four foot tall person. Ordinal measures, in contrast, simply tell us how different quantities rank, but do not say anything about the absolute difference. Suppose I like carrots more than broccoli and broccoli more than brussels sprouts. It would be wrong to conclude from this that I like carrots twice as much as broccoli and three times as much as brussels sprouts. The ordinal ranking provided is consistent with having a slight preference for carrots over broccoli and absolutely loathing brussels sprouts. There is no reason to suppose that the person who came first in the 100 meter dash was twice as fast as the second person, and so on.

In claiming that BIG increases GNF, there is no commitment to the idea that we are X percent freer. All we need for the argument is that such redistribution increases the overall freedom. Just as we can say that the Woodruffs are freer after winning the lottery without being committed to the claim that they are (say) 56.7 percent freer, so too can we claim that BIG will increase GNF without claiming that it will increase GNF by (say) 56.7 percent.

The Intrinsic Value of Freedom: The Brave New World

One point we have yet to address is the value of freedom. We have assumed thus far that freedom has intrinsic value. As noted at the beginning of the chapter, utilitarians disagree with this claim. They maintain freedom is valuable only to the extent that it contributes to happiness. Certainly the utilitarian position has some initial plausibility. If people are freed from bonds of serfdom, it is easy to imagine they are happier. This shows that freedom has instrumental value, at least in some cases. However, utilitarians maintain a much stronger position: freedom only has instrumental value. That is, according to the utilitarian, if freedom does not make us happier, then freedom has no value.

One way to investigate whether utilitarians are correct is to consider the lives of the characters in Aldous Huxley's *Brave New World* (BNW). The novel is often described as a "dystopian future," but, from a utilitarian perspective, this claim is false. Whether happiness is conceived of in terms of life satisfaction or positive affect, the average citizen in BNW is much happier than the average citizen in our world. For example, Ed Diener reports an average American response of 7.73 on a 1–10 life satisfaction scale[33] (on this scale 1 signals "dissatisfied" and 10 "satisfied"). If the same survey were done in BNW, the average would no doubt be very close to 10.

Why are they so happy in BNW? In part, the answer is that material inequality and violence have been solved in a manner one would expect if utilitarian values ruled the world. Wars have long been abolished and no one lacks for food or shelter in BNW. Much of the happiness of denizens of the BNW is a result of technology used to fashion a happy population. The political leaders, "the Controllers," have a cradle-to-grave plan for maximizing happiness. Children are raised in ectogenesis chambers and then in state nurseries. The idea is to break any natural parental bonds and influences that might take away from aggregate happiness. Children are indoctrinated from an early age to be fully satisfied with their lot in society. The controllers work hard to make sure that no long-term pair bonds form. We are told that sexual relations are a source of pleasure and happiness, but strong emotional attachments, which inevitably develop into exclusive relationships, are a source of much unhappiness. Love is replaced by a lifetime of shallow one-night stands. Also banished from BNW are higher forms of culture, serious art, literature, music, philosophy, religion, and science are absent. The Controllers find that these activities negatively influence happiness.

According to its architects, social stability is crucial to the happiness of BNW society, which is maintained in part by the strict caste system. The Alpha and Beta classes are charged with leadership and the others perform the more simple-minded jobs. The Epsilon class is described as having a level of intelligence more akin to that of chimps, so they are assigned tasks like elevator operator. Every level of intelligence is carefully designed to make sure workers are fit for their station and their duties. For some, this is achieved by using alcohol to stunt the future mental capacities of embryos when they are in the ectogenesis chambers. We are told that the caste system is necessary, for if everyone were bright, no one would be happy performing the more menial tasks in society. And so this makes for a perfect fit, the castes with lower intelligence have no ability and no desire to perform the more intellectual tasks. Each caste is conditioned to see its place in society as an enviable one. So the Betas, for example, are told as part of their indoctrination that the lower castes are inferior, while the lower castes are told they are lucky not to be Alphas because the Alphas work too hard. In effect, the argument of the novel is that a stunted intelligence is not a big deal since even the lowest classes, the Deltas and the Epsilons, are happy.

Happiness is also promoted by the use of *soma*, a technically advanced pharmaceutical agent. Huxley introduces *soma* as having "all the advantages of Christianity and alcohol; none of their defects."[34] With *soma* you can "take a holiday from reality whenever you like, and come back without so much as a headache or a mythology."[35]

Thus the puzzle is one that utilitarians must face: why, if happiness is the sole intrinsic value, and BNW is so much happier than ours, do we find BNW so repellent? In his review of *Brave New World*, Bertrand Russell found himself puzzling over just this question. The material prosperity and happiness of BNW is obvious, yet:

> In spite of these merits, the world which Mr. Huxley portrays is such as to arouse disgust in every normal reader, and obviously in Mr. Huxley himself. I have been asking myself why, and trying hard to think that his well-regulated world would really be an improvement upon the one in which we live. At moments I can make myself think this, but I can never make myself feel it. The feeling of revulsion against a well-ordered world has various sources: one of these is that we do not value happiness as much as we sometimes think we do.[36]

Russell's suggestion is quite poignant, "we do not value happiness as much as we sometimes think we do." As Russell is well aware, it is not simply that they are happy in BNW, it is that they are *much* happier. If happiness were the sole intrinsic prudential value, then remaking our world to be like BNW where there is so much happiness ought to be a no-brainer.

So, one way to answer the question of why we do not think BNW is an improvement upon our own is simply to reject utilitarianism. The argument is that if utilitarianism is correct, then the BNW is a better world than ours. Since the BNW is not better, utilitarianism must be false. We noted above that perfectionism is the traditional opponent of hedonism and so of utilitarianism as well. Recall that perfectionism says what is important for a life to go well is the development of mind, character, and body.[37] Of particular interest in the political realm is the idea of autonomy, which is often thought of highly by perfectionists.[38] Autonomy is also something we value as a society, we hope that citizens are able to be the authors of their own lives. And parents who do not encourage the future autonomy of their children are subject to strong social and state disapproval.[39] It is clear that the citizens of the BNW have far less autonomy than we do. Many citizens have diminished rationality because of the alcohol poisoning they suffered as fetuses, which undermines the competency condition of autonomy mentioned above. And since all citizens undergo brainwashing their entire lives, citizens lack autonomy because the authenticity condition is violated by the Controllers. So, although the citizens of BNW are much happier than we are, the price is autonomy.

Since our judgment is that BNW is not a better world than our own, it seems that we must say we are perfectionists, rather than utilitarians. This conclusion, however, is too hasty. There is a third possibility: pluralism. Pluralists say there is more than one intrinsic value. In the present case, we should consider the possibility that both happiness and autonomy are intrinsically valuable. If happiness and autonomy are both intrinsic values, then both utilitarianism and perfectionism are false. In other words, both utilitarianism and perfectionism agree that there is only one intrinsic prudential value, they just disagree on what that value is. Pluralism rejects the one value assumption shared by utilitarianism and perfectionism. So, it is possible to reject the claim that BNW is better than our world without rejecting the idea that happiness is intrinsically valuable. A pluralist might say that BNW is happier than our world, so this counts in its favor, but it has less autonomy, so this counts against the idea that the BNW is better than our world.

So how do we decide between utilitarianism, perfectionism, and pluralism? Here is an easy way to settle the issue. Consider a third world: Mark's Braver New World. This world is equal to our world in terms of autonomy, and equal to BNW in terms of happiness. In Mark's Braver New World, the state does not indoctrinate its citizens or stunt their mental capacities with alcohol like in BNW. Rather, in this world, citizens work together to help each other achieve happiness and autonomy. Instead of being solely focused on happiness like in BNW, both autonomy and happiness are highly valued.

According to utilitarians, which is the best of these three: our world, BNW, or Mark's Braver New World? Answer: BNW and Mark's Braver New World are exactly equal. If a utilitarian had to choose which world to bring into existence, there would be no reason to prefer one over the other. They both have exactly the same amount of happiness. True, Mark's Braver New World has more autonomy, but this is irrelevant to the utilitarian unless it brings about greater happiness. However, by hypothesis, happiness is exactly equal between BNW and Mark's Braver New World. A utilitarian who preferred Mark's Braver New World to the Brave New World would be violating her principles.

According to perfectionists, which is the best of these three worlds? Answer: our world and Mark's Braver New World are exactly equal. If a perfectionist had to choose which world to bring into existence, there would be no reason to prefer one over the other. They both have exactly the same amount of autonomy. True, Mark's Braver New World has more happiness, but this is irrelevant to the perfectionist unless this increased happiness brings about greater perfection. However, by hypothesis, the amount of autonomy is exactly equal between the two worlds. A perfectionist who preferred Mark's Braver New World to the Brave New World would be violating her principles.

According to pluralists, which is the best of these three worlds? Answer: Mark's Braver New World. The reason is that it has more happiness than our world, but not less autonomy, and it has more autonomy than the BNW, but not less happiness. A pluralist who preferred either our world or the Brave New World would be violating her principles.

Notice that the pluralist is not committed to saying that happiness has more value than autonomy or vice versa. The pluralist is committed to saying merely that happiness has at least *some* intrinsic value. Similarly, the pluralist is committed to saying that autonomy has at least *some* intrinsic value.

To put the point in a slightly different manner, think about the inventory and indexing questions of prudential value. The former is the one we have been pursuing: what items are intrinsically prudentially valuable? The latter asks: how much weight should be assigned to the various intrinsic prudential values? These two questions are independent: two people may have the same list of intrinsic prudential values (both agree on the inventory question) but disagree on their relative worth (disagree on the indexing question). For example, I said that a pluralist can say our world has more prudential value than BNW, but such a judgment is not mandated by pluralism itself. Imagine the happiness-first pluralist says, "I believe that happiness and autonomy are intrinsically valuable, but happiness is more valuable than autonomy, this is why BNW is better than our world." The autonomy-first pluralist replies, "I too believe that happiness and autonomy are intrinsically valuable, but autonomy is more valuable than happiness. This is why I believe our world is better than the Brave New World."

Although I have argued for pluralism against perfectionism and utilitarianism, given the long and vexed history of these disputes, it is fortunate that our argument does not require definite adjudication of the inventory and indexing questions. Consider first the inventory question. I argued, on the basis of the three worlds thought experiment, that pluralism is the correct answer to the inventory question: both happiness and freedom (substantive autonomy) have intrinsic value. However, the overall argument is not significantly harmed if either utilitarianism or perfectionism is true. Consider first the case where utilitarianism is correct and pluralism is false. From this assumption it follows that substantive autonomy is not intrinsically valuable. So, the argument of this chapter is not directly supportive of the utilitarian case since, at best, substantive autonomy is instrumentally valuable. Nevertheless, the argument that providing BIG increases GNF does not work against the utilitarian's case. So, even though utilitarians are unmoved by the argument about GNF, utilitarians should be in favor of BIG because of the reasons given in the previous chapter, namely: BIG increases GNH.

Next, consider the case where perfectionism is correct and pluralism is false. From this assumption, it follows that happiness does not have intrinsic value. So, the argument of the previous chapter is not directly supportive of the perfectionist case for BIG, since happiness is, at best, instrumentally valuable. Nevertheless, the argument that providing BIG increases GNP does not work against the perfectionist's brief for BIG. So, perfectionists should be in favor of BIG because of the reasons given in the chapter, namely: BIG increases GNF.

We may summarize the previous two paragraphs more succinctly: utilitarians should get behind chapter 6 and ignore chapter 7. Perfectionists should get behind chapter 7 and ignore chapter 6. Pluralists should endorse both chapters 6 and 7.

Similarly, we do not need to solve the indexing question of which is more valuable, happiness or autonomy. This is because BIG promotes greater happiness *and* autonomy. We would need to say something if the arguments pointed in different directions. For example, if BIG increased happiness but decreased substantive autonomy, something would have to be said about the relative weights of happiness and substantive autonomy. However, since there is no tension in pursuing both of these goals using BIG, there is no pressing need to decide whether happiness is more important than autonomy or vice versa.

Conclusion

It was argued that BIG would contribute to increased GNF. Freedom, properly understood, means something akin to substantive autonomy. We noted too that autonomy is a perfectionist value. It is worth pointing out that BIG may also contribute significantly to many other qualities that perfectionists find valuable. For example, I mentioned in the first chapter that some of my students have suggested BIG might provide for additional educational opportunities. The thought was that graduate school might be a serious possibility if it meant fewer student loans. Other students have suggested following musical careers, pursuing athletic excellence, or choosing entrepreneurial activities using BIG as a backstop to launch such careers. All such activities are valued for their potential to develop human excellence of various kinds. The promotion of such life choices will count as an additional reason for perfectionists and pluralists to endorse BIG.

CHAPTER 8

A BIG Future

Introduction

This chapter responds to some doubts about BIG and looks to the future of political economy. The first objection we will consider is that a rights-based defense of BIG is superior to the consequentialist reasoning for BIG. We will then turn to the charge that BIG encourages parasitism. Along the way we will look to the future of political economy and see that BIG is only a transitory phase as capitalism limps to its death.

There is No Strong Right to BIG

Many friends of BIG may be uncomfortable with the consequentialist orientation of this work. At least since Rawls, there has been a move away from consequentialism in political philosophy. I have not sought to defend consequentialism, but it is worth considering whether a rights-based defense of BIG is inherently superior. One such challenge comes from the worry that an entitlement to BIG seems to be contingent upon good social consequences flowing from BIG. But, so the objection goes, an entitlement to BIG should not be so capricious. We will think of this as the "contingency objection."

Philip Pettit has a nice clear statement for the objection, which he develops in two steps. First, he points out that many proponents of BIG have assumed that an adequate defense of BIG "should establish a claim to a nonnullifiable, nonstigmatizing basic income."[1] As he notes, defenders have typically sought "to make such an income available as a right that is not subject to provisos about existing means, employment history, willingness to perform certain services, or anything of the kind."[2] Pettit maintains further:

> Utilitarian theory makes a very good case for a financially adequate basic income, but it is not clear that it could satisfy the independence

> desideratum. If the government used the utilitarian criterion in making distributional decisions, it might turn out by happy accident that promoting utility would argue for giving each a basic income. But, that would not mean that people would enjoy basic income as a right; they would enjoy it only so long as this was for the utilitarian best.[3]

Pettit uses "utilitarian" in the quote, but the point seems to apply more broadly to the entire consequentialist family, of which utilitarianism is but one member; if circumstances change, then there may no longer be consequentialist reasons for backing BIG. So, the provision of BIG is merely contingent, given consequentialist reasoning.

Pettit seems right about at least this much, consequentialists are forced to say that, under some circumstances, BIG would be the wrong policy for a community to adopt. However, I take this to be a virtue, not a vice, of consequentialist reasoning. The crucial problem is Pettit's argument for the independence condition. Let us assume that Pettit is correct that proponents of BIG have typically sought to establish a right to BIG that is independent of good social consequences. But this is merely to note that many defenses of BIG are not made in consequentialist terms. Unsurprisingly, then, consequentialists will reject the idea that there are political claims that might be established entirely independent of good social consequences. But then such a reply amounts to no more than an appeal to popularity; we should adopt the non-consequentialist independence condition because most theorists in support of BIG are not consequentialists.

Rejecting the appeal to popularity still leaves the question: How can we determine who is correct here? As Pettit notes, there is a good consequentialist case at the moment for BIG. This means that both consequentialists and those who hope to secure BIG as a right cannot appeal to our present situation to adjudicate the disagreement; since both views give the same verdict, we ought to have BIG. We have to look to situations other than our contemporary sociopolitical environment to see whether consequentialists or rights proponents have a more plausible view.

Here is a more extreme situation. Imagine a small farming community that is barely scratching out an existence in the year 999 CE. A charismatic religious leader convinces 15 percent of the population that the world will end in the year 1000 and that only the most devout will be saved. So the 15 percent of the population plan to spend the summer and fall praying rather than farming, with the expectation that the world will end just after winter starts. They assert a right to a BIG, which, if granted, will get them through to the beginning of winter and the world's end. The 85 percent rightly protests that if the 15 percent does not work, not enough of the land will be planted and harvested, and so the community will not have enough food to survive the winter. Many in the community, perhaps all, will die before the first spring crops are harvested. For the good of the community, no BIG should be provided to the 15 percent, and everyone should be forced to work.

Should we side with the assertion of a right to BIG in this case, or with the consequentialist reasoning that BIG should not be provided because it would lead to much unnecessary suffering? Surely the answer is that in an extreme situation like this, BIG should not be granted. But this demonstrates an insurmountable dilemma for proponents of the independence condition. Either we are speaking about a situation relevantly similar to our own, where there is material plenty, or we are talking about an extreme situation where a community has little or no material surplus. If the former, then rights theorists and consequentialists agree that the community ought to provide BIG. If we are speaking about an extreme or emergency situation, then only consequentialism gives the correct verdict that BIG should not be provided.

It is important to note here that the term "rights" is used in a very strong sense. We noted this above in chapter 4 when consequentialist reasoning was contrasted with deontological reasoning. Recall that the consequentialist will switch the train track to kill one person to save five, whereas the deontological theorist, who recognizes a right to life, will not. There is a weaker sense of "rights," however, that consequentialists can endorse. For example, consequentialists might endorse a constitution that guarantees a number of legal rights to citizens.[4] So, for example, consequentialists may support a legal right to freedom of expression even while realizing that in specific situations, this might lead to less aggregate utility. For example, if freedom of expression means that a racist can spout vile nonsense that upsets a large number of people, it might be thought that the consequentialist would be forced to say that the racist should be silenced. However, the consequentialist has a reply to this; in guaranteeing a legal right to freedom of speech, the long-term consequences are much better than the alternative, where there is heavy-handed censorship and no legal right to freedom of expression.

So, there is no inconsistency in consequentialists endorsing legal rights while denying rights in the stronger sense. Thus, a good case could be made for endorsing BIG as a legal right in a society like ours. Indeed, the reasoning of the previous three chapters suggests that there are good reasons for consequentialists to support a constitutional amendment to guarantee BIG as a right along with other rights, such as the rights to freedom of religion and freedom of speech.

Perhaps the most telling reason to doubt there is a strong right to BIG, as opposed to a legal right, is that the right does not appear to be "universalizable." By "universalizable," I mean a right that everyone exercises at the same time. The problem is that only so many can exercise the unconditional aspect of BIG, that is, only so many can live off BIG and refuse all work. What if too many refuse to work, such that there are insufficient resources to provide BIG?

One possible answer from proponents of the strong right conception of BIG is that "this would never happen" because empirical studies confirm that BIG has little effect on the labor participation rate.[5] A few studies show a very small drop in labor participation rate, a few show no drop, and a few indicate a higher rate. So, even in the worst-case scenario, there is no need to worry about such extreme scenarios.

There are several replies to this objection. First, and most important, it does not answer the objection, it merely avoids it. It is like saying that one will never face a situation where one has to decide whether to have a train run over one person to save five, so there is no need to answer the question. True, but that does not tell us whether people have an absolute right to life or whether it is better to sacrifice one for the sake of five.

Second, it is not clear that the studies provide sufficient data to support the idea that "this would never happen." It does not seem entirely implausible that a right to BIG might be used as a political weapon. Consider, for example, the social unrest caused by the Vietnam War, or police violence against racial minorities. Those with progressive sensibilities might welcome the leverage BIG could provide in such cases. But we can also imagine it being used in the name of less progressive causes; for example, people refusing to work until a black president is removed, or until same-sex marriage is outlawed. We can also imagine both sides of a debate attempting to use BIG as a means to leverage an advantage, for example, people both for and against same-sex marriage refusing to work until their view becomes public policy. My point is *not* a prediction that BIG will be used as a political weapon in this way, only that the limited experimental trials using BIG provide us with little relevant data. BIG has never been deployed at a level that is relevant to assessing how likely it is that some might attempt to use BIG in this way.

I suspect that using BIG as a political weapon is unlikely. Nevertheless, it would be foolish, or worse, not to consider what the appropriate response would be in this situation. Once taken seriously, the same dilemma as above rears its head. If BIG is asserted as a right in the strong sense, then even if bad social consequences ensue, we would be obligated to observe this right as far as possible. We would have to continue to honor the right as long as possible—up to the moment the economy collapses. If the right to BIG is not as stringent as this, if it is suggested, for example, that the right might be revoked or suspended in emergency-type situations, then the difference to consequentialism is not obvious. For as we noted above, if the social political situation is much as it is now, then both strong rights proponents and consequentialists may agree on supporting BIG. If the social situation changes radically, then a strong right no longer looks plausible.

In any event, the argument by Pettit is entirely unconvincing on this point. To be convincing, we need some account of rights that diverges from consequentialism in its policy recommendation with respect to BIG. To merely assert, as Pettit does, that there is a divergence, and where there is a divergence it favors the strong rights proponents, is not the least bit compelling. It can easily be countered with the claim that there is a divergence and the divergence favors consequentialism. In either case, we should demand proof. As indicated, in the instances we have considered where there is a divergence, such as in the farming community, the strong rights-based understanding of BIG gives the wrong verdict.[6]

Karl Widerquist has a more nuanced position than Pettit. In general, Widerquist favors a right of noninterference, which includes the right not to join economic projects. BIG, on this view, is a means to honor the right of all those who would prefer not to cooperate economically. He allows that it might be permissible to force people to cooperate in social projects under extreme circumstances. He imagines the following case:

> A rowboat is caught in a swift current above a high falls. Unless everyone on board rows with sufficient strength to pull their own weight, the boat will go over the falls and everyone on board will surely die.[7]

In such a case, Widerquist argues, everyone has a duty to row. The analogous case with BIG is where the good of the community requires that everyone work (pull their own oar). So, Widerquist can say about the farming community that the community is not obliged to support the religious sect that wants to live on BIG largesse, since the survival of the whole community is at stake (the community is at risk of "going over the falls"). According to Widerquist, this differs significantly from the case where the boat is past the point of danger and the majority in the boat would prefer to land upstream. In this case, there is no duty to row (no duty to participate economically when the community's very existence is not at stake). What this shows is that Widerquist's more nuanced position can provide the same verdict as the consequentialist: there is a duty to participate in economic life in the farming community but not in contemporary society. But now we are faced with the same problem. Since Widerquist's theory of the right of noninterference coincides with the consequentialist's view on these two cases, there is nothing to decide which view is correct.

A third example will help us differentiate the views. Imagine once the point of safety is reached, the only landing spot without the help of the reluctant rower is a rocky island in the river. It is determined that the community can survive there, but it will be an extremely difficult existence in perpetuity. If everyone rows for three more hours, they can land upstream where they initially will face a similarly difficult period, but over time the area, once tamed, will offer a much easier existence for the community. The area in question, with some work, will be a farming paradise. The community will eventually be able to spend more time on activities that make them happy and contribute to their excellence. On a consequentialist view, the reluctant rower has a duty to row for another three hours even though the fate of the community is not at stake. The duty is imposed by the fact that there will be so much more long-term happiness and excellence if only the reluctant rower participates for three more hours. According to Widerquist, the reluctant rower has no such duty, even though the long-term prospects of the community are so much brighter if only a few hours of the reluctant rower's time are sacrificed.

This is not the place to decide the issue between consequentialist and rights-based (or deontological) defenses of BIG—this would probably require a book-length treatment itself. The dialectic in this section does reveal two important points. First, the difference between consequentialism and rights-based views about political morality often give the same verdict on specific issues. This observation is further strengthened when we realize that both views have means to incorporate the insights of the other. For example, we noted that the consequentialists can endorse legal or constitutional rights with the realization that extreme circumstances may mean overriding these rights. Rights-based views can incorporate exceptions as well. As we noted, Widerquist's more nuanced view permits exception to the duty of noninterference. Second, every claim about "sacrifice" has a mirror image point from the countervailing view. It will help to illustrate this somewhat turgid claim with reference to earlier examples. One reason for rights-based views of morality is to stop individuals being sacrificed for the social good. As we noted in chapter 4, Nozick goes so far as to say that individuals should never be sacrificed for the social good. If Widerquist is correct, then the right of noninterference should not be overridden unless the fate of others is at stake. This is why the community of rowers would be wrong to cite a duty to contribute by the reluctant rower once the point of safety is reached. The reluctant rower's rights cannot be sacrificed for the sake of the long-term good of society to reach the point up stream, which promises a farming paradise. The mirror image of this point is that society's future happiness and perfection must be sacrificed to ensure that the reluctant rower's rights are honored. The point, then, is that the issue is never one of whether there should be some sacrifice or not. Rather, when consequentialism and rights based understandings of political morality differ on a specific issue, the question is always who should make the sacrifice: the individual or society?

As noted, this is not the place to decide the larger question in normative ethics concerning consequentialism versus deontology. However, I do take it as a point in BIG's favor that BIG can be justified from either starting point.

The Perennial Malibu Surfer

We turn now to what many consider to be the most important objection to BIG: the charge of parasitism. The objection, to recall, is that those who would live off BIG while refusing to look for work would simply live unjustly off the labor of others. Much of the recent literature on this question was spawned by John Rawls, and so it is worth looking at his response in a little more detail.

On the question of material equality, recall that Rawls sought to strike a balance between some versions of socialism, which demand strict equality of wealth, and laissez-faire capitalism, which licenses extreme differences of wealth. So, Rawls says that some material inequality is to be permitted, subject to certain conditions. For our purposes, the most important is that inequalities "are to be of the greatest benefit of the least-advantaged members of society (the difference principle)."[8]

A problem that emerged almost immediately, and has been repeated often since, is that some of the least advantaged members of society might be financially destitute as a matter of choice.[9] As Rawls asks: "Are the least advantaged, then, those who live on welfare and surf all day off Malibu?"[10]

Recall Rawls response to the objection:

> Those who are unwilling to work under conditions where there is much work that needs to be done (I assume that positions and jobs are not scarce or rationed) would have extra leisure stipulated as equal to the index of the least advantaged. So those who surf all day off Malibu must find a way to support themselves and would not be entitled to public funds.[11]

Rawls's basic suggestion, then, is that the least advantaged may take their benefits in different ways. If the least advantaged choose extra leisure—surfing all day—then they cannot claim monetary benefits as well—welfare. This would simply be parasitism.[12] Accordingly, Rawls joins those who make willingness to work a condition for social assistance.[13]

An important point to note here in Rawls's answer to the Malibu surfer objection is that there is an implicit threat made against those who would not work. To say with Rawls that those who do not work will not be entitled to public funds is an oblique way of saying that only the poor will be threatened for idleness. Note that idleness by the rich is acceptable; idleness by the poor is not. This is apparent when we consider the alternative: a universal work imperative. The universal work imperative would force even the rich to work rather than surf all day. Since Rawls does not endorse the universal work imperative, there is a major asymmetry between the rich and the poor: working is optional for the rich, but not for the poor.

While the rich are not required to work in the Rawlsian universe, it seems that Rawls must think it is praiseworthy when they do work. If the rich work, then they will be taxed at a high rate and the funds redistributed to those most in need. Why then is idleness of the rich tolerated? Rawls does not explain his thinking here. Perhaps he thinks that this is simply a consequence of the efficiency emphasis in his difference principle: society should tolerate differences in wealth because allowing inequality is an efficient means to benefit the least well-off. Since society must tolerate a certain amount of inequality, a consequence of this is that the rich may choose to be idle. So, the idleness of the rich is an unintended but foreseen consequence of permitting inequality as an efficient means to help the poor. The idea, perhaps, is that the prospect of not working is part of what is necessary to incentivize the more productive and the more talented to work. It will be argued that a similar line of defense can be made for the idleness of the poor. The idleness of at least some is actually a net positive long-term benefit for society, so it too should be tolerated, if not encouraged. To see our way to this conclusion, we will examine an ubiquitous feature of modern technological economies: threats against idleness.

Threat Economies

In technologically advanced countries, threats are the primary motivator to ensure a dutiful and compliant work force. Although threats in modern industrial economies are ubiquitous, the sources of threats have varied. The Soviet Union required that everyone work. Laws against "parasitism" were often accompanied by hefty sanctions like imprisonment. In terms of practice, such laws were often used against political dissidents, but it is also clear that the laws were on the books to deter free riders: those who might want to take the economic benefits of production without paying the cost in terms of labor. In capitalist countries, the threat attached to not working (for most) is financial destitution. True, unemployment insurance and welfare reduce the threat somewhat. But these are supposed to be temporary or stopgap measures. The United States, for instance, has a five year lifetime limit for federal welfare benefits. These are indirect laws against not working (at least for the poor). Laws against vagrancy and loitering have long been used against the financially destitute.

So, there are strong norms in advanced industrial nations that able-bodied people ought not to live off public assistance, which, in effect, means that there is a strong norm that almost everyone should work. (The case of the idle rich is not an exception to the claim that most should work, since the idle rich are always a minority.)

This leads to *the* contradiction of late-stage capitalism. To understand the contradiction, we need to recall from chapter 5 the relationship between "ought" and "can" statements: "ought" implies "can." The point was made with the following example: Suppose someone says that you ought to save a drowning child. In your defense, you point out that you cannot swim, and even if you could swim, to save the child would mean swimming against a ten mph current. Even Olympic swimmers cannot manage seven mph in the water, so there is no hope. Accordingly, there is no sense in which you *ought* to save the child because you *cannot* save the child. So, to say "ought implies can" means that if you ought to do X (e.g., you ought to save the drowning child), then you can do X (you can save the drowning child). Conversely, if you cannot do X, then it is wrong to say that you ought to do X. The looming contradiction is that we have a strong norm that people *ought to work*, but, as argued in chapter 5, we can foresee that in the coming decades of technological unemployment, it will be the case that many will rightly claim that they *can't find a job*, no matter how hard they try.

Shortening the Workweek

The contradiction could be resolved in a number of ways. If we put a high priority on keeping the poor working, we could simply divide human jobs into ever smaller pieces. For example, I teach six courses a year, so, in theory, my job could be cut in half as part of a national effort to double the number of remaining jobs (assuming my job is among those still performed by humans). If this is not sufficient to keep the poor employed, then my job could be cut into six. The new job description requires each worker to teach one course a year. Obviously, there is a limit to which most jobs can be divided. In terms of policy, this might be done by reducing the workweek. Certainly a ten-hour workweek is imaginable, but a fifteen-minute workweek seems far too impractical. However, at least in theory, shortening the workweek would permit us to keep the threat economy alive.

Let us work an example to see why this is a far from optimal solution from a consequentialist perspective. Suppose there are 80 workers who work 40 hours a week. Technological improvements make half the workers redundant. In effect, the number of employable hours has dropped from 3,200 (80 x 40) per week, to 1,600. Full employment could be restored by cutting the workweek in half: making 20 hours a week the new norm (80 x 20). In terms of generating maximum GNH and GNF, such a policy is suboptimal for two reasons. First, some workers might be happier to pursue non-income-generating ends, for example, climbing mountains or surfing. Second, some workers would prefer to work longer hours and receive more money. So, suppose that 10 percent of the workers would prefer not working and living frugally on BIG, while 90 percent would like to work at least 24 hours a week in order to maximize their earnings. The 10 percent who would prefer not to work translates into 8 workers, so now the 1,600 employable hours is to be divided by 72 workers, meaning that the workweek could be 22 hours rather than 20.

Under these conditions, it is absurd to think that the policy of cutting the workweek to 20 hours and forcing everyone to work (on threat of financial destitution) is a better policy than allowing some to choose BIG and others to choose to work longer hours for more pay. What possible rationale could there be for a policy that makes everyone worse off? The obvious solution for the consequentialist here is to allow some to live off BIG, and others to work longer hours.

Despite his official position against providing the Malibu surfer with BIG, BIG is actually more in tune with Rawls's deeper commitments. To see why, it will help to drill down to look at how the preference about different quantities of income versus leisure noted in the previous example can affect each individual's pursuit of the good life. Consider Sara the surfer, and Freddie the flyer. Suppose once again that technological advances have reduced the need for human labor to the point that in order to ensure full employment, everyone would have to work 20 hours per week. Sara's one passion in life is surfing and she would prefer to receive BIG and not work at formal employment at all. Surfing itself, she reasons, is a very inexpensive activity. Freddie's one passion in life is flying his small plane. It is a very expensive pursuit because of the cost of fuel.

For Sara, BIG is the best option because it maximizes her surfing time. For Freddie, BIG is the best option because it maximizes his flying time. The decision about how to balance income versus leisure is surely an important part of our conception of the good life. Rawls's solution is unresponsive to these differences in conceptions of the good life (table 8.1). Rawls's one-size-fits-all proposal is inconsistent with his deeper commitment to a just society being one where people can cooperate while pursuing their own conception of the good life.

It might be protested on Rawls's behalf that the examples of Sara and Freddie violate one of the assumptions that Rawls makes in his response to the Malibu surfer, namely, "positions and jobs are not scarce or rationed."[14] Unfortunately, the remark is so cryptic and his

Table 8.1 Maximize leisure and income

	BIG	*Rawls's threat*
Sara	0 hours work/$10,000	20 hours work/$30,000
Freddie	40 hours work/$50,000	20 hours work/$30,000
Total	$60,000	$60,000

discussion is carried out at such an abstract level that it is difficult to be fair to Rawls in interpreting this remark.[15] Rather than attempting to parse his meaning, we will argue by dilemma. Either legislative restrictions on the length of the workweek count as rationing, in which case, Rawls should not be against offering BIG to the perennial surfer; or, legislative restrictions on the length of the workweek do not count as job rationing, in which case, threatening those who would prefer not to work is a very unRawlsian solution to the problem of labor participation rates.

The United States has a long history of political control over the labor supply, dating back at least to 1938.[16] These legislative efforts regulate the length of the workweek with hefty monetary penalties for employers who ask their employees to work more than 40 hour per week, that is, with the requirement that they pay time and a half. Indeed, rationing jobs during the Great Depression was one of the reasons Roosevelt introduced the Fair Labor Standards Act. In terms of promoting the social good, it makes sense that the relatively small pool of jobs available during the Great Depression should be divided among as many as possible. It was not uncommon in the period before the legislation was passed for workers to work 50-hour or 60-hour workweeks. Obviously, it makes little sense from a point of aggregate utility to have one worker unemployed and another working 50 to 60 hours a week: better for two workers to work 25 to 30 hours a week. So, if this counts as what Rawls means by "job rationing," then Rawls's objection to the perennial Malibu surfer does not apply in contemporary circumstances. We have had job rationing in many of the rich nations for the better part of a century.

On the other hand, it might be thought that this is not what Rawls meant by job rationing, since it is hard to believe Rawls was unaware of legislative restrictions on the length of the workweek. Yet, if the legislative shortening of the workweek does not count as rationing, then the resulting outcome seems very unRawlsian. As noted, one of the central themes of Rawls's corpus is that society should not impose a conception of the good life on individuals. Society should be organized in such a way that mutual cooperation is possible while allowing maximal scope for individuals to pursue their own conception of what is important in life. It is somewhat baffling just how little attention Rawls gives to the role of work in pursuing the good life in his magisterial, *A Theory of Justice.* The assumption that Rawls seems to employ is that all but the rich will generally work a 40-hour week as part of a "good ordered society." As noted, however, this assumption is inconsistent with people making dramatically different choices about income versus leisure in their lives. Rawls should be an enthusiastic supporter of the perennial Malibu surfer and BIG.

In Defense of the Perennial Malibu Surfer

We are now in a position to summarize the reply to the problem of the perennial Malibu surfer. Rawls's proposal has three features worth commenting on.

First, consider the stipulation that Rawls permits perennial surfers so long as they do not use public money. Notice, however, that the argument of chapter 3 is actually consistent with the public money stipulation. It was argued that one's dividend from state capital is private money in the same sense that money invested in any company is private money. If Warren Buffett can surf all day long and live on the return on his investment, then the rest of us should be entitled to use our return on investment for leisure. It might be protested on Rawls's behalf that the dividend from state capital is actually public money. This response, I believe, is very unRawlsian. One of the fundamental assumptions Rawls shares with Marx is that there is no *natural* divide between public and private when it comes to economic goods. Any divide between public and private is normative, to be decided upon by society as it determines the nature of the just society. Thus, if someone protests against Rawlsian redistribution with the claim, "your taxes rob me of my money," Rawls's reply is that the distinction between what is your money and what is not your money is determined by the distributive justice scheme chosen by a society. If justice requires that there is a progressive tax rate, then the money taken in taxes was not yours to begin with, since it is part of the fundamental agreement by all parties that there will be a progressive tax.

By a similar token, whether we are entitled to a dividend from state capital cannot be simply decided by some appeal to a natural distinction between private and public funds. Rawls says that conventional distinction between private and public funds is to be decided by using the device of a "veil of ignorance." We imagine trying to decide on the fundamental principles governing the operation of society, ignorant of where we actually stand in society. Rawls said that the difference principle is what would be chosen behind the veil of ignorance. That is, wealth is to be redistributed so long as redistribution helps the least well-off. But as we have seen, Rawls faced the perennial surfer objection and so had to modify his theory. Rather than modify the difference principle, Rawls's thought is to force poor surfers to work by claiming that they have more of a primary good, leisure, and so they are not entitled to state funds.

The second point is this: if Rawls is going to endorse an implicit threat against workers—financial destitution if people refuse jobs, then jobs must be rationed even further. As we noted, it is unintelligible to say everyone ought to get a job if there are not enough jobs for everyone. Even now, ignoring the impending robotic revolution, the labor participation rate (a much better indicator of employment than the unemployment rate), shows a trend that will not be reversed in the foreseeable future.[17] If jobs are rationed to bring the unemployment rate down, then the resulting distribution of income/leisure forces a conception of the good life on people in a manner that is unacceptable within Rawls's own strictures of a well-ordered society.

We noted above that people like Sara the surfer and Freddie the flyer might have different visions of the good life that imply different preferences for income and leisure, and Rawls's scheme would result in both of them being worse off. The solution, then, is to modify the distributive scheme to have the difference principle apply to those who work, and allow a basic income for those like Sara who put more emphasis on not working. Not knowing how strong our preference is for work and leisure, this disjunctive principle—either work and have the difference principle apply, or do not work and live on a basic income—is preferable, given Rawls's fundamental assumptions.

Another way to appreciate this point is to see how arbitrary Rawls's assumption is that eight hours of additional leisure should equal the full amount that would be redistributed to the least well-off. Why should eight hours of additional leisure be equal to the full amount rather than some fraction of the full amount? For example, suppose that the market rate for the least advantaged workers working 40 hours a week is $15,000 a year and that, with Rawlsian redistribution, the least economically advantaged workers net $30,000 with redistribution. That is, the least advantaged workers get $15,000 from redistribution in accordance with the difference principle and $15,000 from their employers. Rawls says that the extra leisure of the Malibu surfer should be equivalent to the full $30,000. But why? It appears he is thinking that, otherwise, the Malibu surfer would be entitled to the same $30,000. But this simply does not follow. For a start, half the income from the least well-off among workers comes from the market, so at best it would seem that the Malibu surfer is entitled to $15,000 in redistribution. Rawls's position is that the Malibu surfer is entitled to $15,000 in redistribution only if she works. He does not seem to consider a graduated form of redistribution, for example, one where taking the extra eight hours of leisure reduces by a third what one is entitled to through redistribution. If the Malibu surfer's redistribution is reduced by a third, she would still be entitled to $10,000. So, Rawls seems to be assuming a false dilemma, either the Malibu surfer must work a 40-hour week and is entitled to the full $30,000, or the Malibu surfer is entitled to nothing. But, as we have seen in the case of Sara and Freddie, this is a very unRawlsian view.

The third point is that there are further advantages to perennial surfers not working that Rawls does not seem to consider. One benefit is that for every perennial surfer, the market clearing rate for labor rises. That is, for the US economy as a whole, there is a surplus of workers, there are more people seeking work than there are jobs. Since this point applies to the economy as a whole, it is consistent with shortages of workers in certain segments of the economy. If some drop out of the competitive pool looking for work, this means there will be more competition among employers for employees. Not only should we predict higher wages then, but also better treatment of employees.[18] The present system, where there are not enough jobs for people who would like a job, favors capitalists' interests over workers' interests. If Rawls is serious about helping the least advantaged, he should want to restrict the labor supply in order to drive up wages. As we noted, this could be done either by shortening the workweek or offering BIG, or some combination of both. As things stand at the moment, employers have little incentive to be kind to, or at least treat with a minimum of decency, the least advantaged.

Also of benefit to society is that an increasing scarcity of human labor will speed up the robotic revolution. It is generally acknowledged that there is a certain substitution effect between labor and capital. The exact rate of exchange is disputed but in general; if the price of labor is extremely high, capitalists are motivated to invest in automation to reduce their labor cost. When labor is relatively cheap, then there is less incentive to invest in automation. So, it might not make sense for employers to invest in robotic hamburger flippers when their labor cost is $6 per hour per person, but it may make monetary sense when the labor rate is $10 per hour per person. So, perennial surfers may have the unintended consequence of hastening the robotic revolution that will release many future generations from the drudgery of repetitive jobs. It should be apparent that this argument holds only on two assumptions: first, that those who are most likely to drop out of the work force work in occupations that might be automated, and second, that there is not a lot of withdrawal from the labor market by robotic developers or others that might be instrumental in increasing automation. So, for example, it would be a terrible outcome if BIG induced only those with PhDs in robotics to drop out and surf all day long. Much better for a robotic future is if fast-food workers are more likely to drop out. Given the pay differential and the difference in job satisfaction between roboticists and fast-food workers, it is safe to assume that any decline in the workforce participation rate is more likely to come from the ranks of the latter.

The Rawlsian view that threatens the perennial Malibu surfer is badly mistaken. Perennial surfers ought to be celebrated as heroes of the nation. Without them, the contradiction of late-stage capitalism is only exacerbated.

The Future of Capitalism

The argument for BIG offered here seems to have hitched its fortunes to that of capitalism. In chapter 3, we argued that BIG might be seen as a dividend on shares in state capital. And in chapters 6 and 7, we looked at ways to utilize the productivity of capitalism to generate more GNH and GNF. So, we have plenty of motivation to think about the future of capitalism.

Seemingly relevant in this connection is Thomas Piketty's recent and much discussed *Capital in the Twenty-first Century*. From the title, one might think the book was mostly forward-looking into the prospects of capital and capitalism in the twenty-first century; however, the book is mostly historical. Picketty does a marvelous job tracing the ebb and flow of the accumulation of capital over the past two centuries. He argues on the basis of historical enquiry that income and capital inequalities are the most important economic problems facing us in the twenty-first century. Although he is correct that these are important concerns, he is wrong to think they are the most pressing issues. Picketty mentions, and quickly dismisses, the possibility of advanced robotics radically altering the trajectory of capitalism.[19]

As is perhaps obvious, I am of the opinion that Picketty is mistaken about robotics. Indeed, the situation for capitalism is worse than what has been previously described. It is not simply that technology will cause some technological unemployment. Technology threatens capitalism as we know it. The threat comes from the combination of two forces. The first is one we have previously noted; the observation that the amount of human labor per unit of economic production goes down every year:

> **First Law of Production:** with technological progress, the amount of human labor necessary to create a widget decreases over time.

Widgets are, of course, a catchall for economic goods. We have seen this point several times, for example, in the observation that farm productivity has risen dramatically in the last two centuries, that is, it takes far fewer hours of human labor to produce a bushel of corn today than it did a hundred or two hundred years ago.

The second law is not as widely recognized; the law states that the capital-intensive means of production at one stage of technological development become consumer goods at another stage:

> **Second Law of Production:** with technological progress, the means of production transform from capital goods to consumer goods over time.

To explain the second law, it will help to clarify terminology. The means of production (MOP), it will be recalled, refers to the tools, equipment, machinery, factories, and so on, that are used to produce goods. In capitalist economies, capitalists own the MOP and workers make use of the MOP to create goods. The term "consumer goods" is sometimes defined in terms of items that are ready for consumption and are not used in any further production. Other times, the term "consumer goods" is used to refer to the sorts of things consumers might buy at retail locations. As is perhaps evident, these two definitions are not equivalent. If I buy an oven at a large retail store, the second definition suggests that ovens are consumer goods. However, ovens are used to produce goods, like baked goods or meals, which suggests that they are not consumer goods, since ovens are used to produce other goods. To avoid this confusion, we will think of consumer goods as goods that are not used to produce something for *resale* (at least in the typical case). Thus, although my home oven produces some further produce, the product (dinner, baked goods, etc.) is not intended for resale. That is, the products of my oven are not sold on the open market, so ovens count as consumer goods.

The most important contrast for "consumer goods" is the term "capital goods": goods used in the production of other goods. For example, suppose you start a business making ovens. You will need equipment for your oven-producing factory, for example, metal-fabricating machines, trucks to transport the ovens, computers and desks for office workers to coordinate the sales and distribution of the ovens, and so on. All the tools, machinery, and equipment you buy to set up your factory counts as capital goods. These good are used in the production of consumer goods, the ovens your factory produces, which you hope to sell.

As an example of the second law, think of ice making. At one time, there was a huge industry in North America selling ice to people with iceboxes, and for industrial purposes like making beer and meat packing.[20] Ice was harvested from frozen lakes and stored for the following summer in insulated "ice houses." Some capitalists became very rich off this industry as they owned the MOP that produced ice for consumers; for example, the ice houses where the ice is stored and distributed from. The "ice-man" traveled around neighborhoods selling ice door-to-door. The ice was used then just as we use ice now, in a cooler taken to the beach, directly in drinks (as ice cubes), or to keep foods fresh. Today, most citizens of the United States and other rich nations have a little ice-producing MOP in their own homes, known as a "freezer."

It may seem like a stretch to call a home freezer a MOP. At least part of the reason why we might balk is that we typically think of MOP as a means for producing consumer goods for resale. It would take pretty special circumstances so that I could persuade anyone to buy ice from my freezer. My neighbors have their own freezers. It does not make sense for them to buy ice from me when they can make it at home themselves cheaper (assuming I am trying to make a profit, then I will have no cost advantage over my neighbors). If, through the assistance of a time machine, I could go back to the turn of the nineteenth century carrying with me several freezers (and a generator), I could have made a small fortune selling ice. In this instance, I would be the only one with ice MOP. I would have a cost advantage over my competitors who have to cut, store, and transport ice. However, although we may think in the typical case that MOPs are used to create goods for resale, this is not part of the definition of the MOP. So, there is no contradiction in saying that we live in a time when ice-making MOP (freezers) are consumer goods.

We can see how these two laws often work together. As noted, the amount of human effort to produce a single ice cube has dropped dramatically throughout history. In the days of the ice-man, there was a huge amount of manual labor involved: cutting a frozen lake or stream to harvest blocks of ice, packing the ice in straw, transporting the ice blocks to be stored in ice-houses, and loading ice into wagons to be sold door-to-door. All this labor has been replaced by automated freezers. True, there is still some residual human labor in creating home freezers and maintaining the power grid, but this is a small fraction of the labor that was previously required. However, the two laws are logically independent. Consider, for example, an imaginary world where it is impossible to make ice except by using giant factories the size of football fields. We can suppose the gigantic factory is fully automated and ice is delivered by robotic drivers, but still consumers would be forced to buy from the owner of the MOP. In this case, the law of diminishing labor applies, but not the law of MOP becoming consumer goods.

Another historical example where the MOP becomes a consumer good is in the travel industry. We are witnessing the demise of travel agencies or "travel factories." The value of a travel agency lies primarily in knowledge: knowledge of travel arrangements, destinations, discounts, and so on. At one time, almost all travel arrangements were made through travel agencies, now their share of the market has dwindled to approximately a third (by revenue).[21] Those who owned travel factories, the stored knowledge of the travel agency, were able to make a profit. With the advent of the Internet, the cost of owning a travel factory has drastically dropped. It is possible to look up online prices of different hotels, airlines, cruises, and so on, and also see reviews from customers and others, from the comfort of one's home without paying others for the knowledge. The fact that the knowledge base is now freely available also explains why I will have a hard time trying to make a commission buying airplane tickets for my neighbors. They can just as easily go online and look up the information as I can, so there is little prospect of making a profit.

The printing industry is another example of the second law. Once upon a time, printing was done at a printing shop or "printing factory." Print shop proprietors owned the means of production: printing machines and photocopiers. Formerly, if you lost your dog and wanted to put up flyers featuring a picture of your best friend, you took the original to a printer. Now, of course, almost every home has a printer and a photocopier. The price of printers and photocopiers has plummeted to the point where everyone can be his or her own "printing factory owner" through the ownership of a printer. This also explains why it would be only under very special circumstances that I could make a profit photocopying or printing something for my neighbors. By and large, there is little incentive for them to pay me to do something that they can do just as easily in their own home with their printing MOP.

As we look to the future, we can see that the same force applies to other aspects of the economy. Consider that until very recently it was nearly impossible to obtain electricity in a cost competitive manner except from a large scale producer: one's local utility. We are approaching the crossover point where installing solar panels is more cost advantageous for many homeowners: the MOP (electricity generation equipment) is being commoditized by huge robotic solar panel factories building cheap solar panels. The process is not complete, since most home solar installations simply feed into the central grid. However, the price of batteries continues to drop,[22] so before too long, consumers will be able to go "off grid"—producing all their own electrical energy needs. Individual home owners will own the MOP to produce their own electricity. That is, we will see the rise of millions of "energy factory owners": people producing their own energy from their solar MOP. Again, my neighbors will have little incentive to buy electricity from me when they can buy cheap solar cells and produce their own electricity.

We can foresee too the demise of many fast-food restaurants. We saw in chapter 5 that fast-food robots are on the verge of replacing many fast-food workers. Consider that as fast-food robots get better and cheaper, they will soon be as affordable as today's robotic vacuums and kitchen stoves. At some point, we will be able to have our own fast-food robot right in our own homes. It will merely be a matter of saying "make a cheeseburger and fries for me," and the home fast-food robot will take care of the rest. Again, when a fast-food robot is present in every home, it will be hard for any individual to make a profit selling fast food. If it costs me 30 cents to make a hamburger at home with my fast-food robot, my neighbor is not going to pay me (say) 40 cents for a hamburger when he can make the same hamburger for 30 cents.

We noted that the two laws are not logically related. It is possible that the first law is true and the second law false. However, history bears out the fact that the two laws go hand in glove. The first law of diminishing human labor is specified in terms of the production of widgets. Widgets can be either capital goods or consumer goods. The first law is indiscriminate, it tells us that the amount of labor to produce all goods, consumer and capital, falls over time. Other things being equal, reduced human labor leads to a prediction of lower cost. Given this, it is inevitable that the price of capital goods will fall to the point where the average consumer can afford them, in which case they will no longer be capital goods but consumer goods. Take the ice-making example again. The end of the "ice-man" industry did not happen immediately with the invention of freezers. Early freezers were too expensive for most consumers. Early freezers were capital goods, and capitalists owned them as part of the MOP. For example, restaurants and brewers were early adopters of the technology. They owned freezers as part of their MOP to produce consumer goods, restaurant meals and brew. As they dropped in price, freezers turned from capital goods to consumer goods. However, freezers are MOP in our sense of the term, since they are used to produce a product, ice.

The first law and second law, then, spell trouble for capitalism. Individual capitalists are forced to take advantage of the first law or not survive in the market. As noted, the business of collecting frozen ice from lakes and rivers was relatively labor intensive. The development of freezers reduced the labor and cost of ice. Think about the early days when commercial freezers first became available. Imagine Company X is the old established company that collects ice and stores it in ice-houses. The young upstart, Company Y, uses freezers to make ice. Since Company Y's costs are lower, it sells its ice cheaper and starts to take away market share from Company X. Company X in turn invests in new capital equipment, freezers, and now is able to compete once again with Company Y. Of course, this means that more freezers are now being sold. As the cost of freezers comes down, it becomes possible within a few decades for most people in industrially advanced nations to buy their own ice MOP, and the ice industry suffers a massive die off. (There is a small industry for bagged ice but the glory days of the ice industry are over.)

Therefore, in any particular industry, capitalists are in a bind. On one hand, they must take advantage of the first law or face extinction. That is, they must look for more efficient ways to buy and run their MOP or be out-competed in the market. On the other, they want to avoid the cost of the MOP getting too low, or they face extinction when their MOP becomes consumer goods. A good situation for ice-making capitalists would have been when it was always too expensive to make freezers for the home market. And for print shops, the ideal would have been when it was always too expensive to buy a printer for home use. Of course, the first law tells us that this will not happen. So, as capitalists attempt to reduce the cost of the MOP as a means to stay ahead of their competition, they are hastening the ultimate demise of their particular industry.

If we take this line of thought to its logical conclusion, it seems that we must imagine a world where all the MOP are consumer goods. If this comes to pass, then the free market as we know it will disappear. In the next section, we will examine one such possible future.

The American Dream: 0.4 Acres and a Nanobot

To the suggestion that we are approaching a world where everyone produces all of his or her own goods, it will be remarked that history in fact shows the opposite. It was only a few centuries ago that there was comparatively little economic exchange. Most of what was produced and consumed was on one's own farm. The vision—that production will take place at home—is not a view to the future, but to the past. As Adam Smith long ago noted, it was specialization and trade that sowed the seeds from which capitalism sprung.[23] With increased productivity came new goods and services, which in turn encouraged further specialization and trade. And the examples above seem to support this point. Yes, the ice-man industry vanished, but in its place grew a whole new industry, the manufacture of freezers. While travel agencies may have diminished, the sales of personal computers that allow people to make their own travel arrangements have expanded, from almost no computer ownership thirty years ago to the point where most in the rich nations own several.

There is no denying that the economy has evolved along the lines suggested by Adam Smith: increasingly specialized producers more frequently trade in a competitive market. As usual, the dispute is not about the past—the dispute is about the future. The question is whether the past is a good indicator of what the future will be like, or, whether we have good reason to suppose that things will be radically different. It is the latter claim I want to support.

We live in a world that is much like a giant Lego box. All material things are made up of elements from the periodic table from just under a hundred different types of naturally occurring atoms. Many of the elements are familiar: hydrogen, helium, carbon, and iron. Some are more exotic sounding: cesium and antimony. This means that every material thing is made up of combinations from this basic set of building blocks. Sometimes when I put the key in the ignition of my 1997 rusted Ford minivan and hope it will start, I lament not owning a brand new Porsche convertible. However, I often try to offset this with the happy thought that I actually own all the pieces of a brand new Porsche sports car. After all, a Porsche is basically just a collection of atoms. Atoms of the same type that go into the brand new Porsche in the dealer's showroom can be found in the soil in my backyard. The trouble, of course, is that the atoms are dispersed throughout the soil, perhaps hundreds or thousands of feet down. In theory, however, all the makings of the Porsche are there. It would be unimaginably expensive, but if I had the funds, I could have scientists and engineers separate the iron and the other molecules necessary to recreate a Porsche from the soil in my yard. Here is how (again, in theory) such an operation could go. As a new Porsche rolls off the assembly line in Germany, scientists scan the car down to the molecular level, taking an inventory of each molecule and its relative position to other molecules. The inventory of atoms would be relatively straightforward. It would look something like this: X number of hydrogen atoms, Y number of iron atoms, Z number of carbon atoms, and so on for all the elements on the periodic table. (For some elements, the value may be zero or close to zero.) The basic inventory of atoms then could be printed off on a single page. Much more complex would be the information encoding the relative position of each atom and electron in the Porsche. The instructions for this would be horrendously complex, as anyone who has studied the atomic composition of even single molecules in high school chemistry will know, but with such information, one would have everything one needs to build a Porsche. In effect, the information would constitute a molecular blue print for the Porsche. Suppose then, scientists sift through the dirt in my backyard, looking for the atoms on the inventory list. Once the scientists have amassed all the atoms on the inventory list, they assemble them according to the instructions specifying their relative position to one another. It would be possible in this case to manufacture a brand new Porsche simply from the atomic elements that could be mined from the dirt in my backyard.

Rather than having human scientists and engineers perform the labor, we can imagine nanobots, molecular scale robots, micrometer range in size, collecting molecules and assembling them according to the molecular blueprint. The number of atoms involved is probably somewhere in the order of a number followed by 30 zeros. It would make sense then to have an army of nanobots assigned to the task. Since nanobots themselves are comprised of atoms, it would make sense to assign the first nanobot the task of creating more nanobots and the new nanobots to create further copies of themselves. As is perhaps obvious, in short order, an army of nanobots could be made through the replication process. As with any self-replication process, the first few generations are not that impressive: generation 1 produces 2 nanobots; generation 2, 4 nanobots; generation 3, 8 nanobots; and by the 10th generation, we have just over 1,000 nanobots (1,024). By the twentieth generation, there are over a million nanobots (1,048,576); and by the 30th generation, there are over a billion (1,073,741,824).

Applying this to political economy, consider this future: at age 18, every person is allotted 0.4 acres (about the size of the average middle class single family lot at present) and a nanobot.[24] Imagine you have just turned 18 in this future. You have chosen a piece of land out in the desert far from others, as you value your solitude. Your parents drop you off with a knapsack and your nanobot on the remote property, and, as is customary, leave you to fend for yourself. Using your cell phone, you scan the Internet for molecular blueprints for homes and find one to your liking. You set the nanobot the task of building a house in the most efficient manner possible. Naturally, the nanobot first goes into replication mode to build an army of workers. Each generation takes approximately 20 minutes,[25] you do not see much activity for the first couple of days with your naked eyes, but you do receive updates through wireless communication that the nanobots are successfully replicating. Within a few days, work on your palatial new home is near completion. Again, using schematics obtained on the Internet, you furnish your new home with paintings, appliances, and entertainment systems to your satisfaction. The raw materials for all your needs are found in the ground of your 0.4 acres. Even food is provided straight from the ground. To celebrate your new home, you look over food recipe websites. You decide on a New York style pizza and a bottle of red wine for your celebration. The molecular blue prints for both are provided for the nanobots, which quickly assemble the meal and drink.

I am not suggesting that the nanobot revolution will happen tomorrow—so please do not quit your job just yet. However, it is something like this future we are headed to (assuming we do not blow ourselves up in the meantime, discussed later). How soon until we have technology of this sophistication? Optimists like Ray Kurzweil think it could be somewhere in the order of 2030 to 2040.[26] Whether this is accurate or whether Kurzweil is off by half a century or more is not something we can investigate here. Rather, I would like to make a few points to show that it is not pure fantasy.

The first point is that a precursor technology, 3-D printing, or additive manufacturing, is already in use and rapidly expanding. For those not familiar with the technology, the analogy with a regular printer for printing documents is perhaps most helpful. Think about how inkjet printers put words on a page. A computer program controls an inkjet that deposits tiny amounts of ink according to patterns of letters and symbols. Inkjet printers make a plain sheet of paper a tiny bit thicker as the ink is deposited on a page. 3-D printers work on the same principle with a couple of key differences. First, rather than ink, typically plastic is extruded from a nozzle. Second, numerous layers are laid one on top of another until the object is complete. To make a plastic cup, for example, one simply loads in a design and the printer will print a cup one layer at a time. Many 3-D designs for items, including cups, are available for free from websites like www.thingiverse.com.

The price of 3-D printers has fallen dramatically in the last few years, while the number of items that can be printed by 3-D printer grows every year. Printers that costed over $20,000 less than a decade ago now sell for about $1,000. Some of the items printed using plastic, as the primary input, include prosthetics, shoes, clothing, swimwear, drone helicopters, guitars, and guns. A number of food items are now being 3-D printed, including chocolate and pizza.

3-D printing in the home is still in its infancy. Leading the way is Reprap.org: a community or "open source" movement to develop a free home 3-D printer. More than half the parts of the current generation of RepRap 3-D printers can be printed by one of the printers themselves. The goal is to make a 3-D printer that can print all of its own parts. Work on this goal is being spearheaded by academics and hobbyists across the globe. A study that testifies to just how far 3-D printing has come already was recently made by a group of engineers who found that home 3-D printing is already cost-effective: "A family using one RepRap to print only 20 domestic products per year (about 0.02 percent of the products available) can expect to save between $300 and $2,000."[27] Table 8.2 gives you some idea of how the authors reached their conclusion. They looked up the retail prices of various items that can be made by a RepRap. Retail costs are given for both the lowest cost and the highest cost the authors found for each item.

The cost of making items with a RepRap was less than half, even when using the lowest retail price offered. This is explained in part by the inexpensive inputs into a RepRap: electricity and plastic filament. As can be seen below, the most electricity-intensive item to make, the paper towel holder, cost only 4 cents in electricity. The cost of the plastic filament is relatively inexpensive and set to decline even further. In part, this will be because it may be possible to recycle plastic to use as feedstock. A future RepRap recycling machine might take used plastic milk cartons and repurpose them into plastic feedstock to make something like a paper towel holder or an iPhone dock.

Table 8.2 Cost of some common consumer goods: Retail versus 3-D printing

Product	*Cost of plastic filament*	*Cost of electricity*	*Total RepRap cost*	*Retail cost (Low)*	*Retail cost (High)*	*Savings using RepRap (Compared with low retail cost)*	*Savings using RepRap (compared with high retail cost)*
iPhone 5 dock	$1.62	$0.03	$1.65	$3.56	$29.99	$1.91	$28.34
Jewelry organizer	$0.69	$0.01	$0.70	$9.00	$104.48	$8.30	$103.78
Shower curtain rings	$1.18	$0.03	$1.21	$2.99	$2.99	$1.78	$1.78
Safety razor	$0.35	$0.01	$0.36	$17.00	$78.00	$16.64	$77.64
Paper towel holder	$2.22	$0.04	$2.26	$11.20	$25.00	$8.94	$22.74
Garlic press	$1.58	$0.03	$1.61	$5.22	$10.25	$3.61	$8.64

The engineers conducting this study summarize their most important finding thus:

> As both upgrades and the components that are most likely to wear out in the RepRap can be printed and thus the lifetime of the distributing manufacturing can be substantially increased the unavoidable conclusion from this study is that the RepRap is an economically attractive investment for the average US household already. It appears clear that as RepRaps improve in reliability, continue to decline in cost and both the number and assumed utility of open-source designs continues growing exponentially, open-source 3-D printers will become a mass-market mechatronic device.[28]

Here is another example of how far we have come already: a Chinese company, WinSun, used 3-D printing to print whole houses for about $5,000 each. The material used in the printer is a composite of recycled construction waste, glass, fiber, and cement.[29]

The future is uncertain, but smart money is on the trajectory where 3-D printing becomes cheaper and more capable. It should become cheaper as the printers get the ability to print more of their own components, and as feedstock becomes cheaper through recycling and the ability to use more materials, including dirt.[30] 3-D printers are pretty close to the universal MOP. In theory, just about any material good that you would want could be made with a 3-D printer. We probably will not be able to economically print all the material goods that we may want in ten years time, but almost certainly we will within a hundred.

Neither Capitalism nor Communism

In what follows, I will continue to discuss as if nanobots are likely to be developed. As we have just seen, it would be possible, in principle, to do much of what nanobot technology might be capable, simply at a larger scale. For example, 3-D printers work at a much larger scale than nanobots, but in theory they could be improved to the point where they might be able to print just about anything. In terms of efficiency and flexibility, it is clear that nanobots are to be preferred over macroscopic 3-D printing, but in theory, much of what might be done with nanotechnology could be done with macroscopic 3-D printing robots and other robots to secure raw materials as input for 3-D printer bots. As we noted above in chapter 5, there is already robotic mining equipment. Such macroscopic robots could be put to use mining and refining materials to make useable inputs for 3-D printers. I make this point simply to note that the argument does not hinge on any single technological development. I think the case overwhelmingly favors the idea that not only is nanobot technology possible, it is also likely. Anyone who objects should imagine what follows, redescribed in terms of the activities of 3-D printers and other macroscopic robots.

The nanobot future described is one where all MOP are consumer goods. With one's own supply of nanobots, there is no need for large scale factories owned by a capitalist class oppressing workers; all manufacturing can be right at the point of consumption. It is easy to suppose that, to our descendants, our present economy will look barbaric. Imagine their astonishment to think that someone in the early part of the twenty-first century might have bought furniture made in China with US lumber at a US retailer. Surely this would seem absurd: shipping the molecules (the raw lumber) to China and then shipping the molecules back (in the form of furniture). Of course, at present, China supplies human labor at a lower price that more than covers the cost of shipping, but on a purely molecular level, this seems absurdly inefficient. Compare that to the future where nanobots make a new couch using molecules from an old couch. This is so much more efficient. Not only does it save on shipping the molecules, but it also reduces the cost of buying and disposing of the old couch.

Clearly, nanobots will not solve all problems of distributive justice, but it will permit at least this much: material abundance for all. By "material abundance" I mean something like the material wealth of at least the present upper middle class: a large home, energy, food, and transportation. All these things, as indicated, could be made by nanobots requiring no input of human labor, and no precious natural resources, just dirt or seawater from which to obtain the molecules necessary to assemble the goods. There will be some residual problems of distributive justice associated with some forms of scarcity. For example, there is only one Mona Lisa and a nanotechnology future will not change this. And there is only so much beach front property, and so there will be questions about how this is to be distributed. However, in both cases, a nanotechnology future might reduce these problems. Consider that nanobots would make it possible to own a molecule-for-molecule identical copy of the Mona Lisa: a "fake" so good that no expert could distinguish it from the original. And surfers in a landlocked state like New Mexico need not be deprived of their joy, nanobots could create a saltwater lake in the desert with a wave machine that would guarantee perfect surf every day. After all, even a large saltwater lake is nothing, ultimately, but molecules in motion.

Is such a future a victory for capitalism or communism? As we noted above, capitalism and communism agree that human labor is necessary for production. The residual disagreement between them is how much value human labor ought to be assigned. Capitalists believe that it has market value only, Marx believed that the surplus product extracted by capitalists also belongs rightly to workers and is therefore more valuable. In a nano future, where human labor is no longer required for material abundance, the question of victory looks nearly impossible to answer. On one hand, communists might claim victory since there is no longer exploitation of labor by the capitalist class. On the other, since everyone owns the means of production, capitalists can claim victory—the means of production were not seized by the proletariats, as Marx predicted. In the envisioned future, the working class is absorbed into the capitalist class, as everyone owns their own means of production. In which case, Marx is wrong: the means of production are not owned socially at the end of capitalism. Perhaps, in the spirit of diplomacy, we ought to consider it a tie, since both capitalism and communism achieve their main aim: the MOP remain privately owned and workers are no longer exploited.

Perhaps it might be protested that the victory crown really ought to go to capitalism, since, as noted, it is workers who are absorbed into the capitalist class and not vice versa. However, it should be clear that any victory that the capitalist might enjoy is entirely pyrrhic. Recall that we said capitalism is the conjunction of two theses: private ownership of the means of production and free markets for the buying and selling of goods and services. In the nano future described, the free market for the material goods that make up material abundance has disappeared, since everyone is producing their own material abundance rather than buying and selling with one another. In such a future, I can no more sell a car, created by my nanobots, to my neighbor than I can now sell my neighbor an ice cube.

Indeed, as intimated, the envisioned future overcomes one of the inherent inefficiencies in a market economy: transaction costs. It takes time and effort to buy and sell goods. Suppose today I offer my neighbors free ice and ice chests. I try to sell the idea that they could get rid of their freezers and the associated costs of running them. An energy efficient fridge/freezer might cost \$5 to \$10 a month to run, so this would be a savings for my neighbors. However, there are certain transaction costs involved. They would have to come over and pick up the ice. Even if I offer free delivery, they would have to be home to receive the ice. For most, this would not be worth the hassle to save a few dollars a month. As the price of electricity drops, even getting free ice from me would be a losing proposition: my neighbors could make ice virtually for free and have no transaction costs involved.

It might be objected that there will still be a free market for "high forms" of intellectual labor. Consider, for example, the intellectual labor of writing a novel, writing a song, or making a movie. If robots and computers cannot emulate these forms of human labor, then there is still something that humans can offer the economy. Since the high forms of intellectual labor are scarce, there is every reason to suppose that there will be a free market where these forms of labor will be bought and sold.

There are a couple of problems with this suggestion. The first is that it is not clear that this actually counts as residual capitalism. Yes, there is the buying and selling of goods in a free market, but in this instance, the profit is made on a scarce supply of labor talent, rather than owning the MOP. Furthermore, it is not clear what the untalented will have to trade in exchange. That is, if you do not have the talents that, by hypothesis, are rare, then you do not have anything to exchange. Your labor has been replaced by robotic labor and everyone owns the MOP for material abundance. True, the talented might trade among themselves: you might write the great novel and exchange it for an album of songs. Perhaps such a market might be sustained. However, at least some of those with rare talents will share their works for free. If you and I both write a great novel and I release mine for free while you charge for yours, my audience will be much larger than yours (since by hypothesis only those with rare talents can afford your work). But now I will have something that you might want; a large and adoring audience. Since I do not have to worry about selling my novel to ensure material abundance, there will be a strong pull to release it for fame alone. True, I will not be able to trade my novel for the album that you traded for, but presumably there might also be music that is available free. It is difficult to know in advance, but I suspect that the drive for recognition will overcome the urge to try to maintain a market based on restricted access to the products of those with talents.

In some ways, a future of material abundance is even better than the future predicted by Marx. Marx still labored (as it were) under nineteenth-century conceptions of the MOP. As we noted, his paradigm examples are large factories billowing smoke. Although Marx understood the first law of production, he missed the second. He seems to have imagined large factories that were socially owned. Many of his critics have attacked him on precisely this point: even if factories are not owned by capitalists, this hardly solves all the problems of production and distribution. Some decisions have to be made about how to coordinate production and how the costs and benefits are to be distributed. Marx, as a good student of Hegel, thought that it was impossible to see into the future as to how these problems would be best resolved.[31] To Marx's critics, this seems like an admission that he has no solution. In the nanobot future, coordination problems mostly disappear. We do not have to decide how many hours each of us should work in a giant factory, and whether the factory should produce guns or butter, because there are no factories, and human labor is no longer needed. Nor do we have to decide socially how much each factory should produce, since we decide for ourselves about what we would like our MOP to produce.

A BIG Stepping Stone

If we can foresee the end of capitalism, then does not this also threaten BIG? After all, the argument has been that the enormous fruits of capitalism should be redistributed in order to ensure peace, and to serve the aims of maximizing GNH and GNF. BIG, then, seems beholden to capitalism. If there is no free market, then there are no exchanges from which a transaction fee (that is, tax) can be levied. And if there is no transaction fee, then there will be no funds to pay for BIG. Thus, any thought that capitalism might fade from the political-economic landscape is also a thought about the demise of BIG.

While this may seem a devastating problem for the present proposal, in fact it is not. The reason is that the need for BIG diminishes as the ability to pay for BIG diminishes. As we saw above, there is no need for BIG when we have advanced nanotechnology: 0.4 acres and a nanobot will supply material abundance for all. As we approach this future, the cost of goods will fall because of increasing automation. Consider again the following example of the cost of food over the past two centuries. Most of the time subsistence farmers spent laboring was devoted to producing enough food for their families. Once mechanized machinery was used on farms, the potential for huge surpluses was made possible. Even as recently as 1960, food costs were about 18 percent of disposable income, compared to 10 percent at present.[32] With robotics, there is every reason to suppose that this trend will continue. For example, an average person might eat $60 to

$75 of fresh fruits and vegetables per month. If one is living on BIG alone, this represents about 8 percent of one's income. Now imagine a hydroponic gardening robot can be purchased for $200. The nutrient solution to run the hydroponic unit and the electricity for the robot and circulating pumps might cost a few dollars per month. So, on this scenario, the hydroponic robot would be paid off with three months savings, at which point, the cost of fresh fruit and vegetables might drop from 8 percent to 1 percent or 2 percent of income. We can see, then, the relationship noted above: if industrial farms go broke because people have robots grow food for them, then the tax base will shrink. But, on the other hand, people will need less income if they do not have to buy fresh produce at their local grocery store. The more the tax base shrinks, the more that will be produced at home, at the point of consumption. BIG, then, is a way for us to transition from a capitalistic mode of production to a postcapitalistic mode of production.

The biggest residual problem we face in transitioning to such an economy concerns what Marx called the "subjects of labor," rather than the "instruments of labor." Marx used the term subjects of labor to include things such as land and natural resources, and instruments of labor to refer to things such as factories and tools that human workers use to produce goods. The instruments of labor, as we have seen, are the sorts of things that will be radically transformed by the coming robotic revolution. The supply of the instruments of labor is not fixed, we can make more factories or tools. However, the subjects of labor do seem fixed. As the old saying goes, "They aren't making any more land." This means that there will still be a problem of distributive justice: how will the finite supply of the subjects of labor be distributed?

While it is true that the subjects of labor are fixed, in a certain sense, it is also true that technology can radically change the value of the subjects of labor. We noted this above in connection with the idea that a piece of desert might be transformed into a palatial residence by nanotechnology. At present, such desert land is virtually worthless: it would require an enormous amount of labor to transform it into something useable. In the nanotechnology future, the labor is provided by technology, hence, this changes the equation. Similarly, think about the relative scarcity of gold. No more gold is being produced and nanotechnology is not likely to change this fact. But consider that there is about a hundred times as much gold in the oceans as compared with all the gold that has been mined over the course of human history. Of course, this gold is virtually worthless now because it is so diluted in seawater it is not economical to separate it out. With advanced nanotechnology, all this changes. The labor and energy to harvest gold from the ocean is basically free: nanobots can provide the labor, and energy can be harvested from the sun.

The point then is not that the subjects of labor are not scarce, but that they are not as scarce as they may first appear. Consider that if every adult citizen in the United States at present was granted 0.4 acres, as imagined above, as part of a homesteading act, this would represent only about 6 percent of all US land. Interestingly, 0.4 acres is far more land than would be required for an individual to produce his or her own food and energy.[33] So, with the victory of the robotic revolution, there should be plenty of land for everyone, so long as it is distributed according to principles of justice. Conversely, if the right to some subject of labor, that is, some natural resource, is not recognized, then there is almost certainly going to be civil unrest, if not outright revolution. Imagine, when the robotic revolution is complete, workers will be unable to sell their labor and have no way to pay rent for land. If the masses are told they can use the land only if they can pay for it, and there is no way to pay for it, then the situation will be ripe for violent revolution.

Techno-Optimism

It may be thought that the argument discussed earlier suffers from a bad case of techno-optimism: the idea that technology will solve all our problems.

This allegation I vehemently deny. I readily admit to arguing that questions of distributive justice evolve with technology, but this is a far cry from the claim that technology will solve all our problems. One reason was already mentioned: even the most advanced technology will not solve some questions of distribution, for example, who gets to own the original Mona Lisa. So, even within the domain of distributive justice, there are problems that can only be solved socially, not technologically.

More importantly, I believe that advanced technology, like 3-D printing and nanotechnology, is a doubled-edged sword. We have talked about some of the positives of such technology in helping with material want. However, one of the downsides of advanced technology is its destructive capacity. For example, we have already seen the first 3-D printing of guns.[34] The weapons are fairly crude, but the threshold has been breached. With advanced robotics, it will be much easier for individuals to make weapons that are much more deadly, unless we build in safeguards. It is easy to imagine, for example, nanotechnology being used to create viruses, to directly attack one's enemy by (say) building blockages in the main arteries, or assembling bombs. In short, technological advancements may make widespread warfare rooted in personal manufacturing of weapons a reality. The problem, in a nutshell, is that warfare is a type of labor. Until recently, a tremendous amount of human labor went into building and wielding weapons. With the robotic revolution, this will no longer be true. So, technology raises the possibility of the complete destruction of civilization and perhaps of the human species.

So, the charge of techno-optimism is mistaken. It should be obvious that BIG is not the cause of these extinction worries, it is advanced technology. If anything, BIG may serve as one possible means to alleviate tension. We noted in chapter 5 that technological unemployment has the potential to cause unrest and that BIG is a good solution to help reduce social tension. The same point applies here: if BIG is in place, it will reduce motivation fueled by financial destitution to use technology for terrorism and war. Obviously, this is not the only thing we should consider to reduce threats from technology, but BIG may go some distance to reducing the threat from advanced technologies.

It is interesting that there is a sort of cyclical nature of political philosophy. Political philosophy in the early Enlightenment was primarily concerned about peace, particularly problems of civil war, and much less about distributive justice. In the nineteenth and twentieth centuries, the primary focus was on distributive justice. In the twenty-first century, political philosophy ought to focus again on peace rather than distributive justice, given the potential of advanced technologies to destroy millions, or perhaps, civilization in its entirety.

CHAPTER 9

Concluding Unscientific Postscript

We Have Been Robbed

To summarize the argument: we have been robbed.

If you believe in an unrestrained capitalism, one that is not tempered by concern for the social good, then you should agree that we have been robbed of our just capitalistic desserts. Citizens in the industrially advanced nations have shares in a huge market place, one that makes eBay pale in comparison, and we do not make sufficient return on our capital. BIG is the least we should expect as shareholders. We have so far been robbed of our dividend.

If you believe that happiness is an important value in governing society, then you should agree that we have been robbed of happiness. As a society, we would be happier if BIG were part of the income distribution scheme of this country. The fact that BIG is not yet a reality means we have been robbed of happiness.

If you believe that freedom is an important value in governing society, then you should agree that we have been robbed of freedom. As a society, we would be freer if BIG were part of the income distribution scheme of this country. The fact that BIG is not yet a reality means we have been robbed of freedom.

Is BIG Merely Utopian

What should we say about the objection that BIG is merely utopian? The objection, in other words, is that as much as BIG might be a good thing, it will never happen. It is politically unworkable.[1]

In some sense, this objection is irrelevant because this is a work in normative politics; it tells us what *should be done*, not whether it *will be done*. Consider a parallel: Thomas Jefferson included antislavery language in the original draft of the Declaration of Independence. The language was edited out by others. It is not hard to imagine that even those sympathetic with Jefferson's view might have thought prohibition against slavery would never happen. But pessimism about reform does not change the fact that it was true slavery ought to have been abolished in 1776, as Jefferson wanted, even though it was not. Similarly, it is true that we ought to institute BIG today, even if society does not agree. This line of thought is not an indictment of the BIG proposal, but of society itself.

Still, this response seems to suggest that BIG is in the realm of "mere academic concern." Interestingly, one reason to think that BIG is not merely of academic interest is because there is something like a precursor to BIG already at work in the United States, namely, the Alaska Permanent Fund, mentioned earlier. The fund was set up in 1976 with the purpose of providing the citizens of Alaska dividends from the state's oil revenues. The fund typically pays in the order of $1,000 to $2,000 per year per person. It is a precursor since the dividend is nowhere near sufficient to live on, especially in a state that has a higher cost of living than many other part of the United States. Still, the idea is not entirely foreign to American politics.[2]

Moreover, one of the lessons we learn from history is that progressive social changes are very hard to predict. Same-sex marriage is permitted in many US states in 2015, yet, in 1976, the Democratic convention refused to pass a resolution that recognized homosexuals as human beings. Very few would have predicted in 1856 that the centuries old practice of slavery would end within ten years. [3] If someone had asked me five years ago to rank the country that would first permit the sale of marijuana in stores, I would have ranked the United States last on a list that included Britain, France, Canada, Australia, and New Zealand. Yet, in two US states, Washington and Colorado, people are free to buy pot in retail locations as I write this. The point then is that those who claim BIG is utopian, claim to know too much: it is extremely difficult to predict when progressive changes will take place. Any number of events can precipitate change. Some of these are large-scale events, for example, the Civil War and World War I were catalysts for the end of slavery and enfranchising women. Other times, a single person's actions, like those of Rosa Parks, can excite many to action and act as a catalyst for change.

History teaches both negative and positive lessons. The positive lesson one can draw here is that change almost never happens unless some people agitate for it. Race relations in the United States are not what they should be, but there can hardly be any doubt that things would be worse but for those like Martin Luther King and Rosa Parks, who agitated and sacrificed for change, knowing that their cause was righteous. This work has tried to make the case that there are good moral arguments for BIG. We are entitled to BIG as a dividend of our share of state capital, and BIG will serve the interests of peace while increasing GNH and GNF. This tells us that the cause is righteous.

The negative lesson is that there is no way to predict with any certainty when change will take place. There is no history oracle, and no crystal ball. There is no privileged vantage point to predict the march of the world spirit. Agitation for change in the face of such uncertainty is always a leap of faith.

The question going forward is not whether we should have BIG, but how to turn our moral insight, that we should have BIG, into political action. The next step involves understanding strategy: how to effect the political change necessary to make BIG a reality. In this respect, I am an amateur, and so I leave to those more experienced in political advocacy the task of charting a way forward.[4]

(The book should have ended with the previous paragraph, but I cannot resist weighing in on the question of strategy despite my amateur standing. The most direct way forward is to form a political party with BIG as its central message. I will discuss the US situation with the understanding that each country has its own political history that might require different approaches.

It is not necessary to subscribe to the opinion that there is no difference between the Democratic and Republican parties to say that, in practice, they are not too far apart. Those who are knowledgeable about the politics of other developed nations understand that, against such a backdrop, the Republican and Democratic parties are both right-wing parties. Similarly, both parties, for whatever reason, have proven to be stalwarts against progressive changes, at least since the Regan era.

What is needed is a party that looks to the future, rather than clings to the past. Consider the Happiness and Freedom Party (HFP) that has BIG as its central plank, along with universal health care, and other measures that are designed to increase the gross national happiness and freedom of the nation. One thing that offers a ray of hope is that, under the present system, candidates do not need a majority to win an election. In theory, under the present system, a candidate in a three-way race can win with 34 percent of the votes when the other two candidates have 33 percent each. Less optimistically, even winning a small percentage of the votes would be enough to get BIG to be part of the national conversation, especially if the party acts as a spoiler: taking enough votes from either party to cause a change in the election outcome, and hopefully, a change in their platforms. The HFP asks for your vote.)

Notes

1 Basic Income Guarantee

1. Eric Alden Smith et al., "Production Systems, Inheritance, and Inequality in Premodern Societies," *Current Anthropology* 51, 1 (2010):8 5–94.
2. Allen Sheahen suggests the Bible as a precursor to contemporary thinking about BIG. Allan Sheahen, *Basic Income Guarantee: Your Right to Economic Security* (New York: Palgrave Macmillan, 2012). For a brief history of BIG, see the very informative page by the Unconditional Basic Income Europe: http://ubie.org/brief-history-basic-income-ideas/. For a survey of early American thinking on this matter, see Jamie Bronstein, "A History of the BIG Idea: Winstanley, Paine, Skidmore and Bellamy," *Journal of Evolution & Technology* 24, 1 (2014), http://jetpress.org/v24/bronstein.htm.
3. The perennial Malibu surfer is a familiar figure from the Basic Income literature of John Rawls, "The Priority of Right and Ideas of the Good," *Philosophy & Public Affairs* 17, 4 (1988): 251–276. We will discuss the surfer more later.
4. Karl Widerquist (perhaps very wisely) does not get too much into specifics in his great work on BIG, Karl Widerquist, *Independence, Propertylessness, and Basic Income: A Theory of Freedom as the Power to Say No* (New York: Palgrave Macmillan, 2013). On the other hand, Allan Sheahen delves into the nitty-gritty in his wonderful recent book. Sheahen, *Basic Income Guarantee: Your Right to Economic Security.* My discussion on the practical details owes much to Sheahen's work. Perhaps the most ambitious proposal is made by Robert R. Schutz, *The $30,000 Solution: A Guaranteed Annual Income for Every American* (Santa Barbara, CA: Fithian Press, 1996). As Karl Widerquist points out, this is quite the guarantee, given that the average salary at the time (1996) was a couple of thousand dollars less than the proposed amount. (Karl Widerquist, "Perspectives on the Guaranteed Income, Part I," *Journal of Economic Issues* 35, 3 (2001): 749–757.)
5. Bradley R. Schiller et al., *The Economics of Poverty and Discrimination* (Upper Saddle River, NJ: Prentice Hall, 2001). Stuart White, "Social Minimum," 2004, http://stanford.library.usyd.edu.au/entries/social-minimum/.

6. "2012 HHS Poverty Guidelines," *U.S. Department of Health and Human Services*, accessed May 10, 2012, http://aspe.hhs.gov/poverty/12poverty.shtml.
7. Data from "Living Wage Calculation for Las Cruces City, Dona Ana County, New Mexico," *Living Wage Calculator*, accessed May 10, 2015, http://livingwage.mit.edu/places/3501339380.
8. There is some controversy as to whether Milton was for a basic income guarantee (through a negative income tax) or whether he was arguing conditionally: *If* you believe society should provide welfare payments to the poor, then a negative income tax is an efficient system that promotes freedom. Milton Friedman, *Capitalism and Freedom* (Chicago: University of Chicago Press, 1962).
9. R. Nozick, *Anarchy, State and Utopia* (New York: Basic Books, 1974).
10. Bertrand Russell, *Roads to Freedom: Socialism, Anarchism and Syndicalism* (London: Allen & Unwin, 1966).
11. Erik Olin Wright, *Envisioning Real Utopias* (London: Verso, 2010).
12. Kai Nielsen, *Equality and Liberty: A Defense of Radical Egalitarianism* (New Jersey: Rowman & Littlefield, 1984). Gerald Allan Cohen, *Why Not Socialism?* (Princeton, NJ: Princeton University Press, 2009).
13. Andrew P. Wilper et al., "Health Insurance and Mortality in US Adults," *American Journal of Public Health* 99, 12 (2009).
14. Widerquist, *Independence, Propertylessness, and Basic Income*,1 10.
15. Rawls rejects the "welfare state" in his later work, John Rawls, *Justice as Fairness: A Restatement* (Harvard, MA: Harvard University Press, 2001). His argument is that the welfare state does not go far enough in insuring that the least advantaged are treated justly in terms of access to the material goods of society. He uses "welfare state" in a slightly different sense in his work from what we are employing. The main point of invoking Rawls here is to cite a theorist who wants to keep some semblance of a free market but use the power of the state to ensure an outcome closer to an egalitarian distribution than typically results from laissez-faire capitalism.
16. J. Rawls, *A Theory of Justice* (Harvard, MA: Harvard University Press, 1971).
17. Ibid.
18. Rawls, "The Priority of Right and Ideas of the Good."
19. Rawls, *A Theory of Justice.*
20. Simon Birnbaum, *Basic Income Reconsidered: Social Justice, Liberalism, and the Demands of Equality* (New York: Palgrave Macmillan, 2012).
21. Stuart White, *The Civic Minimum: On the Rights and Obligations of Economic Citizenship* (Oxford, UK: Oxford University Press, 2003).
22. Lawrence C. Becker, "The Obligation to Work," *Ethics* (1980): 35–49.

23. Widerquist, *Independence, Propertylessness, and Basic Income.*
24. Representative examples include Philippe van Parijs's "real freedom," Philippe van Parijs, "Real Freedom for All: What (if Anything) Can Justify Capitalism?" (Oxford: Oxford University Press, 1995); and Philip Pettit's "republican freedom," Philip Pettit, "A Republican Right to Basic Income?," *Basic Income Studies* 2, 2 (2007): 1–8. I would say this is true even of Birnbaum's Rawlsian inspired defense of BIG. Birnbaum argues that state neutrality on the question of the good life requires BIG, which, I believe, can be understood as a point about the scope of our freedom to choose the good life. Birnbaum, *Basic Income Reconsidered: Social Justice, Liberalism, and the Demands of Equality.*
25. A notable exception to the rights orientation of much work on BIG is Robert Goodin's utilitarian defense of BIG as a means to promote aggregate happiness, Robert E. Goodin, *Utilitarianism as a Public Philosophy* (New York: Cambridge University Press, 1995).
26. Thad Williamson, "Who Owns What? An Egalitarian Interpretation of John Rawls's Idea of a Property-Owning Democracy," *Journal of Social Philosophy* 40, 3 (2009): 434–453.

2 Paying for Basic Income Guarantee

1. Stan Collender, "Stop And Smell The Roses: Final 2014 Federal Deficit Fell . . . Big Time," *Forbes*, October 16, 2014, http://www.forbes.com/sites/stancollender/2014/10/16/stop-and-smell-the-roses-final-2014-federal-deficit-fell-big-time/.
2. B. E. A. US Department of Commerce, accessed May 11, 2015, http://www.bea.gov/newsreleases/national/gdp/gdpnewsrelease.htm.
3. The 44 million figure comes from "State & County Quick Facts," *The United States Census Bureau*, April 29, 2015. The point about social security being analogous to BIG comes from Allan Sheahen (Sheahen, *Basic Income Guarantee: Your Right to Economic Security* (New York: Palgrave Macmillan, 2012), 87).
4. The estimate is based on Sheahen's work. Sheahen's calculation is based on 36 million seniors at a cost of 40 billion. Since we now have 44 million seniors, I have boosted the top-up amount accordingly. (See Sheahen, *Basic Income Guarantee: Your Right to Economic Security*, for Sheahen's numbers.)
5. Adam Liptak, "1 in 100 US Adults behind Bars, New Study Says," *New York Times* 28 (2008), http://sites.google.com/a/trammell-criminology.com/www/NewYorkTimesArticle.pdf.
6. Jeffrey Reiman and Paul Leighton, *The Rich Get Richer and the Poor Get Prison*, 10th ed. (New York: Pearson, 2012).

7. The estimate is based on a total of 13.3 million immigrants (Nancy Rytina, *Homeland Security Office of Immigration Statistics*, n.d., http://www.dhs.gov/sites/default/files/publications/ois_lpr_pe_2012.pdf.) less children, seniors, prisoners, and immigrants, who are already excluded from our analysis. The conservative assumption is that children and seniors make up the same proportion of immigrants as the general population. I say this is a conservative assumption since children born to immigrants in this country themselves are not immigrants. So, in all likelihood, there are more than 8.3 million immigrants in the 18–64 age range. On the other hand, at least some of the 8.3 million are eligible to become citizens, and so with financial incentive, the number of immigrants may be reduced. Not that it matters, but in the interest of full disclosure, I should make clear that I am a permanent resident (that is, immigrant) in the United States at the time of writing this (November 2014).
8. There is a relative paucity of literature on BIG and children. The US BIG network has over 2,000 books and articles in its database on BIG, but using their search function returns only six titles dealing specifically with children. Most of the emphasis is on the possibility of using BIG as a means to encourage children to go to school in the developing nations, for example, see Eduardo M. Suplicy and Christovam Buarque, "A Guaranteed Minimum Income to Eradicate Poverty and Help Poor Children Go to School Instead of Being Forced to Work. The Brazilian Debate and Experience," in *BIEN's 6th Congress, Vienna*, 1996, 12–14. A more recent and popular piece dealing with the United States argues that BIG might not work for adults, but should be considered for children, Clio Chang, "Give Families With Children Cash to Help Them Out of Poverty—US News," *US News & World Report*, November 5, 2014, http://www.usnews.com/opinion/economic-intelligence/2014/11/05/give-families-with-children-cash-to-help-them-out-of-poverty.
9. A. Sheahen, *It's Time to Think BIG! How to Simplify the Tax Code and Provide Every American With a Basic Income Guarantee* (USBIG Discussion Paper, 2006), http://usbig.net/papers/144-Sheahen-RefundableTaxCredit.pdf.
10. Any surplus should be invested to help in case of potential revenue drops in times of recession.
11. OECD, "Value Added Taxes Yield, Rates and Structure," in *Consumption Tax Trends* (Organisation for Economic Co-operation and Development, 2008), 39–71, http://www.oecd-ilibrary.org/content/chapter/ctt-2008-4-en.
12. "Detailed National accounts/GDP Statistics for More than 100 Countries—Charts, Time Series, Historical Data," *Bluenomics*, accessed May 11, 2015, https://www.bluenomics.com/data#!data/national_accounts_gdp/gdp_expenditure_approach/structure_

of_gdp_expenditure_approach/structure_of_gdp_expenditure_approach_annual_nominal_of_gdp|chart/line&countries=usa&sorting=list//title.

13. US Census Bureau, "Current Population Survey," *US Census Bureau*, January 14, 2013, http://www.census.gov/hhes/www/cpstables/032011/perinc/new01_001.htm.
14. The New York Times, "Effective Income Tax Rates," 2012, http://www.nytimes.com/interactive/2012/01/18/us/effective-income-tax-rates.html?_r=2&.
15. Thomas Piketty, *Capital in the Twenty-First Century* (Massachusetts: Harvard University Press, 2014).
16. Allan Sheahen (1932–2013) was a longtime proponent of BIG. We follow here Sheahen's last major proposal: Sheahen, *Basic Income Guarantee: Your Right to Economic Security.* Many of Sheahen's calculations can be found in Appendix A (153–160). Many of the figures date from 2010. I have mostly used these figures. As noted, the economy has improved considerably since 2010, so things are actually a bit better than described in most of this section.
17. Ibid.,8 8.
18. "Key Issues: Tax Expenditures," *Government Accountability Office*, accessed May 11, 2015, http://www.gao.gov/key_issues/tax_expenditures/issue_summary.
19. Adding to the national debt simply means that it is paid in the future by someone else.
20. Sheahen, *Basic Income Guarantee: Your Right to Economic Security.*
21. "Annual Inflation Adjustments for 2013," *Internal Revenue Service*, January 11, 2013, http://www.irs.gov/uac/Newsroom/Annual-Inflation-Adjustments-for-2013.
22. \$10,000 BIG + \$6,200 – \$930.
23. Sheahen, *Basic Income Guarantee: Your Right to Economic Security*, 89.
24. Ibid. See appendix A for details on some of the federal programs and how these should be reduced.
25. Sam Perlo-Freeman et al., *Trends in World Military Expenditure, 2012* (Sweden: SIPRI, 2013).
26. Ibid.
27. "Rudy Turns Up the Heat—The Rush Limbaugh Show," *Rush Limbaugh*, April 25, 2007, http://www.rushlimbaugh.com/daily/2007/04/25/rudy_turns_up_the_heat.
28. Andrew J. Bacevich, *The New American Militarism: How Americans Are Seduced by War* (New York: Oxford University Press, 2013).
29. Ron Paul, "Imagine," *From Reason to Freedom*, March 9, 2009, http://www.reasontofreedom.com/node/1775.
30. Annie Lowrey, "C.B.O. Cuts 2013 Deficit Estimate by 24%," *The New York Times*, May 14, 2013, http://www.nytimes.com/2013/05/15/business/cbo-cuts-2013-deficit-estimate-by-24-percent.html.

31. Harvey Morris, "U.S. Health Care Costs More Than 'Socialized' European Medicine," *IHT Rendezvous*, accessed May 12, 2015, http://rendezvous.blogs.nytimes.com/2012/06/28/u-s-health-care-costs-more-than-socialized-european-medicine/.
32. "Health as a Share of GDP, 2000–2011, Selected G7 Countries," *OECD Library*, October 31, 2013, http://www.keepeek.com/Digital-Asset-Management/oecd/social-issues-migration-health/health-at-a-glance-2013/health-expenditure-as-a-share-of-gdp-2000-11-selected-g7-countries_health_glance-2013-graph153-en#page1.
33. Ibid.
34. Ellen Nolte and C. Martin McKee, "In Amenable Mortality—Deaths Avoidable through Health Care—Progress in the US Lags That of Three European Countries," *Health Affairs* (2012), 2114–2122.

3 Fulltime Capitalism: Basic Income Guarantee as a Dividend from State Capital

1. Jean-Jacques Rousseau, *The Social Contract* (Penguin, 1971).
2. "Capitalism," *Merriam-Webster*, accessed May 12, 2015, http://www.merriam-webster.com/dictionary/capitalism.
3. Richard A. Epstein, "Against Redress," *Daedalus*, 2002, 39–48.
4. This sort of criticism is explored very capably in J. Reiman, *As Free and as Just as Possible: The Theory of Marxian Liberalism* (UK: Wiley-Blackwell, 2012).
5. Ting Wang, Ling Wang, and Tiffany Chen, "Secrets of Success: Innovation in the Silicon Valley: Business Development and Innovation of Tech Companies," in *Behavior, Economic and Social Computing (BESC), 2014 International Conference on* (IEEE, 2014), 1–5, http://ieeexplore.ieee.org/xpls/abs_all.jsp?arnumber=7059530.
6. I am lumping all public property under the federal banner simply for convenience. Municipal parks and state universities would also count as part of the assets of US citizens.
7. US Census Bureau Demographic Internet, "Current Population Survey."
8. The resulting distribution might be more like the proposal to provide citizens a "stake" upon reaching adulthood. Bruce Ackerman and Anne Alstott, *The Stakeholder Society* (Yale University Press, 1999).
9. For more on the mildly paternalistic reasons for not providing lump sums, see Parijs, "Real Freedom for All: What (if Anything) Can Justify Capitalism?" *OUP Catalogue*,1 997.
10. Sheahen proposes a 2 percent wealth tax, which raises $450 billion (Sheahen, *Basic Income Guarantee: Your Right to Economic Security*, 92.) We did not help ourselves to revenue from this tax in chapter 2 in anticipation of this point. Piketty also sees an inheritance tax as an important measure to reduce inequality. Piketty, *Capital in the Twenty-First Century* (Harvard University Press, 2014).

11. "2014 Land Report 100," The Land Report: The Magazine of the American Landowner (n.d.), http://www.landreport.com/americas-100-largest-landowners/.
12. "Population by Sex, Rate of Population Increase, Surface Area and Density," *United Nations Stats*, 2006, http://unstats.un.org/unsd/demographic/products/dyb/dyb2006/Table03.pdf.
13. One way that U.$. Inc. might do this is to replace at least some of the current income tax with the aforementioned "value-added tax" (VAT). The idea, roughly, is that while it may be difficult to get foreign companies to pay income tax, collecting the VAT and turning over the proceeds to U.$. Inc. is a much easier way to stop such attempts at parasitism.
14. Even a number of libertarians have sympathy with this idea. Cf. Peter Vallentyne, "Libertarianism and the Justice of a Basic Income," *Basic Income Studies* 6, 2 (2011).
15. Thomas Paine, *Agrarian Justice* (Alex Catalogue, Raleigh, NC: W. Adlard,1 974).
16. Karl Widerquist, "The Physical Basis of Voluntary Trade," *Human Rights Review* 11, 1 (2010): 83–103.
17. Widerquist, *Independence, Propertylessness, and Basic Income: A Theory of Freedom as the Power to Say No* (New York: Palgrave Macmillan, 2013).
18. Philippe Van Parijs, "Why Surfers Should Be Fed: The Liberal Case for an Unconditional Basic Income," *Philosophy & Public Affairs* (1991): 101–131; Parijs, "Real Freedom for All."
19. G. Van Donselaar, *The Right to Exploit: Parasitism, Scarcity, and Basic Income* (New York: Oxford University Press, 2009); Gijs Van Donselaar, "The Freedom-Based Account of Solidarity and Basic Income," *Ethical Theory and Moral Practice* 1, 3 (1998): 313–333.
20. Jurgen De Wispelaere, "Sharing Job Resources: Ethical Reflections on the Justification of Basic Income," *Analyse & Kritik* 22, 2 (2000): 237–256; Birnbaum, *Basic Income Reconsidered: Social Justice, Liberalism, and the Demands of Equality* (New York: Palgrave Macmillan, 2012).

4 Capitalism: Consequentialism versus Rights

1. Ashley Parker, "House Republicans Cheer Boehner's Refusal to Negotiate on Cuts," *The New York Times*, February 28, 2013, http://www.nytimes.com/2013/03/01/us/politics/house-republicans-cheer-boehners-refusal-to-negotiate-on-cuts.html.
2. Nozick, *Anarchy, State and Utopia* (New York: Basic Books, 1974).
3. I borrow the camping analogy from Gerald Cohen (Cohen, *Why Not Socialism?* (Princeton, NJ: Princeton University Press, 2009)).
4. Karl Marx and Friedrich Engels, *The Communist Manifesto* (Peking: Foreign Language Press, 1975).

5. Adam Smith, *An Inquiry into the Nature and Causes of the Wealth of Nations* (A. and C. Black, 1863).
6. Ibid.
7. Anne Robert Jacques Turgot and Peter Diderik Groenewegen, *The Economics of ARJ Turgot* (The Hague, Netherlands: Martinus Nijhoff, 1977).
8. Nozick also stipulates that legitimate justice in transfer must be free of fraud and deception.
9. Interestingly, Nozick justifies this in terms of the benefits that capitalism brings: those born later without any land, for example, are better off financially working for a property owner than they would have been otherwise. The irony (or inconsistency) here is that now the justification turns to the net benefits of capitalism. That is, consequentialist reasoning (appeal to the social good) seems to be perfectly acceptable to Nozick at this point in his argument, even though such appeal is not acceptable when it comes to taxes. For more on Nozick's difficulties at this stage of his argument see, Widerquist, *Independence, Propertylessness, and Basic Income: A Theory of Freedom as the Power to Say No* (New York: Palgrave, 2013).
10. Massimo Motta, *Competition Policy: Theory and Practice* (New York: Cambridge University Press, 2004).
11. Patrick Bolton, Joseph F. Brodley, and Michael H. Riordan, "Predatory Pricing: Strategic Theory and Legal Policy," *Georgetown Law Journal* 88 (2000): 22–39.
12. Paul Milgrom, "Predatory Pricing," *The New Palgrave Dictionary of Economics* 3 (1987): 938.
13. Milton Friedman and Rose Friedman, *Free to Choose: A Personal Statement* (New York: Houghton Mifflin Harcourt, 1990), 53.

5 Peace, Robots, and Technological Unemployment

1. This chapter borrows very substantially from a previously published article: Mark Walker, "BIG and Technological Unemployment: Chicken Little Versus the Economists," *Journal of Evolution & Technology* 24, 1 (2014): 5–25.
2. John Markoff, "Skilled Work, Without the Worker," *New York Times*, 2012, http://www.nytimes.com/2012/08/19/business/new-wave-of-adept-robots-is-changing-global-industry.html?pagewanted=all&_r=0.
3. Ibid.
4. Timorthy Lee, "Amazon Envisions Eventually Delivering Packages in 30 Minutes via Drones," *The Washington Post*, 2013, http://www.washingtonpost.com/blogs/the-switch/wp/2013/12/01/amazon-wants-to-deliver-packages-in-30-minutes-with-drones/.
5. Growing a Nation, "Historical Time Line," 2005, http://www.agclassroom.org/gan/timeline/.

6. Ira Sager, "Farm Bots and Two-Month-Old Bread: Innovations in Food and Agriculture," 2013, http://www.businessweek.com/articles/2013-01-10/farm-bots-and-two-month-old-bread-innovations-in-food-and-agriculture. Erik Sofge, "3 New Farm Bots Programmed to Pick, Plant and Drive," *Popular Mechanics*, 2009, http://www.popularmechanics.com/technology/engineering/robots/4328685.
7. Vinod Khosla, "Technology Will Replace 80% of What Doctors Do," *CNN*, 2102, http://tech.fortune.cnn.com/2012/12/04/technology-doctors-khosla/.
8. Ralph Gonzales et al., "Excessive Antibiotic Use for Acute Respiratory Infections in the United States," *Clinical Infectious Diseases* 33, 6 (2001): 757–762.
9. Khosla, "Technology Will Replace 80% of What Doctors Do."
10. J. D. Rockoff, "Robots vs. Anesthesiologists," *Wall Street Journal*, Pg. B8–9, 2013.
11. Paul Krugman, "Degrees and Dollars," *New York Times* 6, March (2011), http://www.lib.tku.edu.tw/service/exampdf/dayexam/01day/10143027.pdf.
12. David Levy, "The Ethics of Robot Prostitutes," in *Robot Ethics: The Ethical and Social Implications of Robotics*, ed. Patrick Lin, Keith Abney, and George A Bekey (Cambridge, MA: MIT Press, 2011), 223–232.
13. Raja Roy, "Exploring the Boundary Conditions of Disruption: Large Firms and New Product Introduction with a Potentially Disruptive Technology in the Industrial Robotics Industry," 2014, http://ieeexplore.ieee.org/xpls/abs_all.jsp?arnumber=6578147.
14. Kevin Kelly, "Better Than Human: Why Robots Will—And Must—Take Our Jobs," *Wired*, 2012, http://www.wired.com/gadgetlab/2012/12/ff-robots-will-take-our-jobs/all/.
15. Ibid.
16. I do not want to brag, but I was once "crew member of the month" at my local McDonald's.
17. Peter Murray, "Robot Serves Up 360 Hamburgers Per Hour," *Singularity Hub*, 2013, http://singularityhub.com/2013/01/22/robot-serves-up-340-hamburgers-per-hour/.
18. Casey Chan, "Sushi Restaurant Uses Sushi Robots and Control Centers to Cut Costs," *Gizmodo*,2 010.
19. The tendency to underestimate the robotic revolution is apparent even in Brynjolfsson and McAfee's wonderful book *Race Against the Machine* (Brynjolfsson and McAfee, *Race against the Machine: How the Digital Revolution Is Accelerating Innovation, Driving Productivity, and Irreversibly Transforming Employment and the Economy* (Lexington, MA: Digital Frontier Press, 2011)). They mention that humans still have a competitive advantage in fine motor skills, and so gardeners and busboys have occupations that are safe,

at least for now. However, the Kura chain demonstrates the dispensability of much of the labor of busboys, and robotic lawn motors are busy today replacing the labor of human gardeners. The most detailed and methodologically sound investigation of the issue I know of can be found in Frey and Osborne (2013). They identify 702 distinct occupations in the US work force and have found that "about 47 percent of total US employment is at risk" Carl Benedikt Frey and Michael A. Osborne, "The Future of Employment: How Susceptible Are Jobs to Computerisation?" *Retrieved September* 7 (2013). http://assets.careerspot.com.au/files/news/The_Future_of_Employment_OMS_Working_Paper_1.pdf.

20. Keith Wagstaff, "How Foxconn's Million-Machine 'Robot Kingdom' Will Change the Face of Manufacturing," *Time*, 2011, http://techland.time.com/2011/11/09/how-foxconns-million-machine-robot-kingdom-will-change-the-face-of-manufacturing/.
21. Bien Perez, "Hon Hai Plan to Deploy 1m Robots Faces Delay, Says Analyst," *South China Morning Post*, 2012, http://www.scmp.com/business/companies/article/1052557/hon-hai-plan-deploy-1m-robots-faces-delay-says-analyst.
22. Henry Blodget, "CEO of Apple Partner Foxconn: 'Managing One Million Animals Gives Me A Headache,'" *Business Insider*, 2012, http://www.businessinsider.com/foxconn-animals-2012-1.
23. M. Walker, "A Moral Paradox in the Creation of Artificial Intelligence: Mary Poppins 3000s of the World Unite!," *AAAI Press* Human Implications of Human-Robot Interaction: Papers from the AAAI Workshop (2006): 23–28.
24. Bureau of Labor Statistics, "Labor Force Statistics from the Current Population Survey," *Department of Labor*, 2010, http://www.bls.gov/cps/cps_htgm.htm.
25. Alex Tabarrok, "Productivity and Unemployment," *Marginal Revolution*, 2003, http://marginalrevolution.com/marginalrevolution/2003/12/productivity_an.html.
26. One worry here is how unemployment is measured. If workers give up looking for work, they do not count as unemployed by official government accounting. To make the unemployment rate drop, then, the nefarious government leaders need not necessarily create jobs, but simply a climate of despair. Some evidence for this increased level of despair is another measure of employment: labor participation rates. The percentage of the population classified as "workers" has trended down since the turn of the century.Also, as Karl Widerquist has pointed out to me, stagnant wages are another indication of the effect of increased productivity. Both lines of evidence require far more discussion than I can provide here. Still, there is reason to believe that official unemployment rate figures are hardly decisive evidence that the gains in productivity have not negatively affected the labor market.

27. Gregory Clark, *A Farewell to Alms: A Brief Economic History of the World* (Princeton, NJ: Princeton University Press, 2008). This passage is quoted in Brynjolfsson and McAfee, *Race against the Machine.*
28. There are still a few specialized uses for horses, such as use by police in parks and crowd control situations.
29. Frey and Osborne, "The Future of Employment."
30. Karl Widerquist and Michael A. Lewis, "An Efficiency Argument for the Basic Income Guarantee," *International Journal of Environment, Workplace and Employment* 2, 1 (2006): 21–43.

6 BIG Happiness

1. I am assuming a version of "welfarism." I believe the argument can incorporate non-welfarist conceptions of "the good," but ignore that here as an unnecessary complication. On the issue of welfarism, see Simon Keller, "Welfarism," *Philosophy Compass* 4, 1 (2009): 82–95.
2. Thomas Hurka, *Perfectionism* (New York: Oxford University Press, 1993). For criticism of perfectionism see L. W. Sumner, "Two Theories of the Good," in *The Good Life and the Human Good*, edited by Ellen Frankel Paul, Fred D. Miller, Jr., and Jeffrey Paul (New York: Cambridge University Press, 1992), 19; and Mark Walker, *Happy-People-Pills for All* (Hoboken, NJ: John Wiley & Sons, 2013).
3. S. Lyubomirsky, L. King, and E. Diener, "The Benefits of Frequent Positive Affect: Does Happiness Lead to Success?," *Psychological Bulletin* 131 (2005): 803–855.
4. The term "emotional state view" comes from Daniel Haybron, "Happiness and Pleasure," *Philosophy and Phenomenological Research* 62, 3 (2001): 501–528; *The Pursuit of Unhappiness* (New York: Oxford University Press, 2008). Haybron does not offer the emotional state view as a monistic conception of happiness. He suggests that happiness is a "mongrel" concept that defies such analyzes.
5. Haybron (D. Haybron, "On Being Happy or Unhappy," *Philosophy and Phenomenological Research* 71, 2 (2005): 287–317.) offers a detailed and slightly different inventory of positive moods and emotions than is typically found in the psychology literature (Cf. Lyubomirsky, King, and Diener, "The Benefits of Frequent Positive Affect: Does Happiness Lead to Success?"). This is one of the "details" that we will hope a sufficiently robust emotional state theory will resolve, but one that is not of immediate concern, since the present argument does not turn on how this question is adjudicated.
6. Lyubomirsky, King, and Diener, "The Benefits of Frequent Positive Affect: Does Happiness Lead to Success?"
7. Haybron, *The Pursuit of Unhappiness.*
8. For more on the question of whether sensory pleasure should be included as part of our understanding of 'happiness', see Walker, *Happy-People-Pills for All* (New York: John Wiley, 2013).

9. A competitor is Fred Feldman's "attitudinal hedonism" account of happiness, F. Feldman, *What Is This Thing Called Happiness?* (New York: Oxford University Press, 2010). We will not consider this view here. I have argued that there is significant overlap between Feldman's account and the whole life satisfaction view, so the contrast is not as stark as Feldman suggests. Walker, *Happy-People-Pills for All.*
10. R. Veenhoven, "Is Happiness Relative?," *Social Indicators Research* 24, 1 (1991): 1–34. R. Veenhoven, "Happiness in Nations," *Subjective Appreciation of Life in* 56 (1993): 1946–1992.
11. E. Diener et al., "The Satisfaction with Life Scale," *Journal of Personality Assessment* 49, 1 (1985): 71–75.
12. W. Tatarkiewicz, *Analysis of Happiness*, vol. 3 (Warszawa, Poland: Kluwer Academic Press, 1976). R. Nozick, *Philosophical Explanations* (Cambridge, MA: Belknap Press, 1981). E. Telfer, *Happiness* (New York: St. Martin's Press, 1980). L. W Sumner, *Welfare, Happiness, and Ethics* (New York: Oxford University Press, 1996).
13. Sumner, *Welfare, Happiness, and Ethics.*145.
14. A number of whole life satisfaction theorists add an affective element. For example, Ibid. 111.
15. Elsewhere I have argued that both components are part of our understanding of happiness and that both are important for living a good life Walker, *Happy-People-Pills for All.*
16. Daniel Kahneman and Angus Deaton, "High Income Improves Evaluation of Life but Not Emotional Well-Being," *Proceedings of the National Academy of Sciences* 107, 38 (2010): 16489–93.
17. Betsey Stevenson and Justin Wolfers, *Subjective Well-Being and Income: Is There Any Evidence of Satiation?* (National Bureau of Economic Research, 2013), http://www.nber.org/papers/w18992.
18. Daniel W. Sacks, Betsey Stevenson, and Justin Wolfers, "The New Stylized Facts about Income and Subjective Well-Being," *Emotion* 12, 6 (2012): 1181–1187.
19. Data are reported by Stevenson and Wolfers, *Subjective Well-Being and Income.*
20. G. Standing, *The Precariat: The New Dangerous Class* (New York: Bloomsbury, 2011).
21. The Gallup Poll is based on household income but BIG is paid individually. Households with more than one adult will have a lower net tax, since they will be eligible for more than one BIG payment.
22. Bruno S. Frey, "Happiness: A Revolution in Economics," *MIT Press Books* 1 (2008). Carol Graham, *Happiness around the World: The Paradox of Happy Peasants and Miserable Millionaires* (New York: Oxford University Press, 2009).
23. The literature on this topic is enormous, but for some representative articles, see Alan B. Krueger and David A. Schkade, "The Reliability of Subjective Well-Being Measures," *Journal of Public*

Economics 92, 8 (2008): 1833–45; Sonja Lyubomirsky and Heidi S. Lepper, "A Measure of Subjective Happiness: Preliminary Reliability and Construct Validation," *Social Indicators Research* 46, 2 (1999): 137–155; Deborah Mattel and Charles Schafer, "An Investigation of Validity of the Subjective Happiness Scale 1," *Psychological Reports* 94, 1 (2004): 288–290; E. Diener, "Subjective Well-Being: The Science of Happiness and a Proposal for a National Index," *American Psychologist* 55, 1 (2000): 34.

24. Lyubomirsky and Lepper, "A Measure of Subjective Happiness."
25. R. A. Easterlin, *Does Economic Growth Improve the Human Lot? Nations and Households in Economic Growth: Essays in Honour of Moses Abramovitz. M. Abramovitz, PA David and MW Reder* (New York; London: Academic Press, 1974).
26. For some accessible history on the Easterlin paradox, see David Leonhardt, "Maybe Money Does Buy Happiness After All," *The New York Times*, April 16, 2008, section Business, http://www.nytimes.com/2008/04/16/business/16leonhardt.html.
27. Easterlin, *Does Economic Growth Improve the Human Lot?*,1 18.
28. Stevenson and Wolfers, *Subjective Well-Being and Income.*
29. Sacks, Stevenson, and Wolfers, "The New Stylized Facts about Income and Subjective Well-Being," 1182.
30. Ibid.p .1 186.
31. This is not quite right since Easterlin says there is some minimal threshold a country must reach before income no longer buys happiness. We can safely ignore this wrinkle here.
32. E. Diener et al., "Dispositional Affect and Job Outcomes," *Social Indicators Research* 59, 3 (2002): 229–259.
33. Lyubomirsky, King, and Diener, "The Benefits of Frequent Positive Affect: Does Happiness Lead to Success?"
34. Ibid.
35. Walker, *Happy-People-Pills for All.*
36. "Per Capita Income by State," *Bureau of Business and Economic Research, UNM*, April 2, 2013, https://bber.unm.edu/econ/us-pci.htm.
37. In this section, I concentrate on the claim by socialists to strive for a more radical equality of income. In the final chapter, I will address the question at the forefront of Marxist critiques of capitalism: workers should own the means of production.
38. Jeffrey Thompson, *Raising Revenue from High-Income Households: Should States Continue to Place the Lowest Tax Rates on Those with the Highest Incomes?* (Political Economy Research Institute, University of Massachusetts at Amherst, 2012), https://ideas.repec.org/p/uma/perips/revenue_peri_march5.html.
39. Frey, "Happiness."
40. An analogous point is made by many authors to the effect that welfare is "stigmatizing," a "visible marker of status." Birnbaum, *Basic*

Income Reconsidered: Social Justice, Liberalism, and the Demands of Equality (New York: Palgrave Macmillan, 2012), 50. See also Catriona McKinnon, "Basic Income, Self-Respect and Reciprocity," *Journal of Applied Philosophy* 20, 2 (2003): 143–158.

41. Giacomo Corneo, "Work Norms and the Welfare State," *CESifo Economic Studies* 58, 4 (December 1, 2012): 599–625, doi:10.1093/cesifo/ifs017.
42. Frey, "Happiness."
43. Widerquist and Lewis, "An Efficiency Argument for the Basic Income Guarantee," *International Journal of Environment, Workplace and Employment* 2, 1 (2006): 21–43.

7 BIG Freedom

1. Cf. Friedman, *Capitalism and Freedom* (Chicago: University of Chicago Press, 1962).
2. "Hoosier Lottery," accessed May 17, 2015, https://www.hoosierlottery.com/.
3. Lee A. Iacocca, Sonny Kleinfield, and Ray McKenzie, *Talking Straight* (New York: Bantam books, 1988), 67. Quoted in Gurdip S. Bakshi and Zhiwu Chen, "The Spirit of Capitalism and Stock-Market Prices," *The American Economic Review*, 1996, 133–157, 133.
4. "Per Capita Income by State."
5. Lawrence Blume and Steven Durlauf, *The New Palgrave Dictionary of Economics* (New York: Palgrave Macmillan, 2008).
6. Mathias Trabandt and Harald Uhlig, "The Laffer Curve Revisited," *Journal of Monetary Economics* 58, 4 (2011): 305–327. Mathias Trabandt and Harald Uhlig, *How Do Laffer Curves Differ across Countries?* (National Bureau of Economic Research, 2012), http://www.nber.org/papers/w17862.
7. Jon Gruber and Emmanuel Saez, "The Elasticity of Taxable Income: Evidence and Implications," *Journal of Public Economics* 84, 1 (2002): 1–32.
8. Chye-Ching Huang, "Recent Studies Find Raising Taxes on High-Income Households Would Not Harm the Economy," *Center on Budget and Policy Priorities*, 2012, http://www.cbpp.org/cms/index.cfm?fa=view&id=3756.
9. "The Distribution of Household Income and Federal Taxes, 2010," *Congressional Budget Office*, accessed May 17, 2015, http://www.cbo.gov/publication/44604.
10. Huang, "Recent Studies Find Raising Taxes on High-Income Households Would Not Harm the Economy."
11. Gerald F. Gaus, *Justificatory Liberalism: An Essay on Epistemology and Political Theory* (New York: Oxford University Press, 1996).
12. Reiman, *As Free and as Just as Possible: The Theory of Marxian Liberalism* (Malden, MA: Wiley-Blackwell, 2012).

13. Thomas Hobbes, *Leviathan: On the Matter, Forme and Power of a Commonwealth and Civil* (New York: Collier, 1962), ch. 14.
14. I am not endorsing this line of thinking; I am merely reporting it. The idea that owning more property than others does not affect one's negative liberty has been challenged in a very compelling fashion by Gerald Cohen (Gerald Allan Cohen, *Self-Ownership, Freedom, and Equality* (Cambridge University Press, 1995)). We will make an argument against negative liberty below.
15. Isaiah Berlin, "Two Concepts of Liberty," in *Four Essays on Liberty* (Oxford: Oxford University Press, 1969), 118–172.
16. Gerald F. Gaus, *Political Concepts and Political Theories* (Boulder, CO: Westview Press, 2000), ch. 5.
17. Thomas Hill Green, *Lectures on the Principles of Political Obligation and Other Writings* (New York: Cambridge University Press, 1986).
18. R. H. Tawney, "Equality (1931)," *Italics in Text* (London: Allen & Unwin, 1952), 241.
19. Ibid.
20. I am simply conceding the point here. As Gerald Cohen has forcefully argued, even on the negative conception of freedom there seems to be a vast difference between the negative freedom of the Woodruffs before and after the lottery winnings (Cohen, *Self-Ownership, Freedom, and Equality*).
21. Friedrich A. Hayek, *The Constitution of Liberty* (Chicago, IL: Chicago University Press, 1960), 17.
22. Isaiah Berlin, *Four Essays on Liberty* (Oxford: Oxford University Press, 1969), 53, quoted in Cohen.
23. John Rawls, *Political Liberalism: Expanded Edition* (New York: Columbia University Press, 2005), 325–326. As Rawls later makes clear, this is simply a stipulative definition.
24. Philip Rucker, "GOP Senate Hopeful in Alaska Preaches 'more Freedom, Less Government,'" *The Washington Post*, October 10, 2014, http://www.washingtonpost.com/politics/senate-hopefuls-say-theyll-show-washington-the-alaskan-way/2014/10/10/7891dd6a-4e97-11e4-babe-e91da079cb8a_story.html.
25. J. Locke, *Two Treatises of Government* (New York: Cambridge University Press, 19908), second Treatise, chapter 6.
26. Christman, John. "Autonomy in moral and political philosophy," *Stanford encyclopedia of philosophy*(2008).
27. J. Christman, "Autonomy," *Stanford Encyclopedia of Philosophy*, 2009.
28. Harry G. Frankfurt, *Freedom of the Will and the Concept of a Person* (Clifton, NJ: Springer, 1988).
29. Marina Oshana, *Personal Autonomy in Society* (Burlington, VT: Ashgate, 2006). Stephen Kershnar, "A Liberal Argument for Slavery," *Journal of Social Philosophy* 34, 4 (2003): 510–536, 513. Joseph Raz, *The Morality of Freedom* (Oxford University Press, 1986).

30. Some have disputed the cogency of such contracts. See Arthur Kuflik, "The Inalienability of Autonomy," *Philosophy & Public Affairs*, 1984, 271–298. For some response, see Kershnar, "A Liberal Argument for Slavery." As for myself, I do not see any way to enforce the prohibition against slave contracts other than an appeal to the aggregate good of society. To see why, consider what Karl Widerquist has to say in this connection: "expanding people's positive opportunities is a lesser priority than protecting their core freedoms from interference. Therefore, the government must not take positive action to enforce unconscionable contracts including those alienating status freedom" (Widerquist, *Independence, Propertylessness, and Basic Income: A Theory of Freedom as the Power to Say No*, 58). The idea, in other words, is that to protect someone's freedom, society must not enforce slave contracts. But Widerquist's response merely papers over the problem: in not permitting such contracts we may sometimes harm the interests of the very individuals we are trying to protect, for example, the husband in the example. A blanket prohibition on slave contracts will have to appeal to something else other than that such a prohibition always serves the interests of the party contracting to be a slave. The most obvious candidate for something other is the good of society.
31. The point that property rights can have profound consequences for freedom even when freedom is negatively conceived is made forcefully by Widerquist, *Independence, Propertylessness, and Basic Income: A Theory of Freedom as the Power to Say No.* As he puts it, "A theory of freedom that takes property rights as given is not a theory of negative freedom," p. 41.
32. Ian Carter, "Positive and Negative Liberty," 2003, http://stanford.library.usyd.edu.au/entries/liberty-positive-negative/.
33. Diener, "Subjective Well-Being: The Science of Happiness and a Proposal for a National Index," *American Psychologist* 55, 1 (2000): 34.
34. A. Huxley, *Brave New World and Brave New World Revisited* (Canada: Vintage Books, 2007).
35. Ibid.
36. B. Russell, "We Don't Want to Be Happy," in *Aldous Huxley*, ed. C. Watt (London, UK: Psychology Press, 1997).
37. Hurka, *Perfectionism.* Steven Wall, *Liberalism, Perfectionism and Restraint* (New York: Cambridge University Press, 1998).
38. Thomas Hurka, "Review: [untitled]," *Mind*, New Series, 110, 439 (July 2001): 878–881.
39. Joel Feinberg, "A Child's Right to an Open Future," in *Whose Child? Children's Rights, Parental Authority and State Power* (New Jersey: Rowman & Littlefield, 1980).

8 A BIG Future

1. Pettit, "A Republican Right to Basic Income?" *Basic Income Studies* 2, 2 (2007): 1–8. Pettit's positive proposal is couched in terms of a republican theory of freedom. For more on the connection between republicanism and BIG, see Antoni Domènech and Daniel Raventós, "Property and Republican Freedom: An Institutional Approach to Basic Income," *Basic Income Studies* 2, 2 (2007). Frank Lovett, "Domination and Distributive Justice," *The Journal of Politics* 71, 03 (2009):8 17–830.
2. Pettit, "A Republican Right to Basic Income?"
3. Ibid.
4. Wayne Sumner, *The Moral Foundation of Rights* (New York: Oxford University Press, 1987).
5. Karl Widerquist, "A Failure to Communicate: What (if Anything) Can We Learn from the Negative Income Tax Experiments?," *The Journal of Socio-Economics* 34, 1 (2005): 49–81.
6. Many rights-based discussions of BIG assume that the social circumstances involve advanced industrial countries, Philippe van Parijs, "Real Freedom for All: What (if Anything) Can Justify Capitalism?" *OUP Catalogue*, 1997. Simon Birnbaum, *Basic Income Reconsidered: Social Justice, Liberalism, and the Demands of Equality* (New York: Palgrave Macmillan, 2012); and Karl Widerquist, *Independence, Propertylessness, and Basic Income: A Theory of Freedom as the Power to Say No* (New York: Palgrave Macmillan, 2013). This suggests that they *reject* the strong rights-based defenses of BIG. But, as noted above, it becomes difficult to say which account is superior, a weak right to BIG or consequentialist reasoning, when both agree that BIG ought to be offered in advanced industrial nations, and both can support a legal right to BIG.
7. Widerquist, *Independence, Propertylessness, and Basic Income: A Theory of Freedom as the Power to Say No*, 174.
8. John Rawls, *Justice as Fairness: A Restatement* (Cambridge, MA: Harvard University Press, 2001).
9. Richard A. Musgrave, "Maximin, Uncertainty, and the Leisure Trade-Off," *The Quarterly Journal of Economics* (1974): 625–632. Will Kymlicka, *Contemporary Political Philosophy* (Oxford: Oxford University Press, 2001), 72–74.
10. Rawls, *Justice as Fairness*,1 79.
11. Ibid.
12. Cf. Van Donselaar, *The Right to Exploit: Parasitism, Scarcity, and Basic Income* (New York: Oxford University Press, 2009).
13. Stuart White, *The Civic Minimum: On the Rights and Obligations of Economic Citizenship* (New York: Oxford University Press, 2003). Elizabeth S. Anderson, "What Is the Point of Equality?," *Ethics* 109, 2 (1999): 287–337.

14. John Rawls, *Political Liberalism: Expanded Edition* (New York: Columbia University Press, 2005), 182, n. 9.
15. Even Birnbaum's sustained examination of BIG and Rawlsian liberalism does not address this point. Birnbaum, *Basic Income Reconsidered: Social Justice, Liberalism, and the Demands of Equality.*
16. David R. Roediger and Philip Sheldon Foner, *Our Own Time: A History of American Labor and the Working Day* (London: Verso, 1989).
17. Edward P. Lazear and James R. Spletzer, *The United States Labor Market: Status Quo or a New Normal?* (National Bureau of Economic Research,2 012).
18. Widerquist, *Independence, Propertylessness, and Basic Income: A Theory of Freedom as the Power to Say No*,1 05.
19. Thomas Piketty, *Capital in the Twenty-First Century* (Cambridge, MA: Harvard University Press, 2014), 217.
20. Andrew Hind, "Brrrr! History of the Ice Industry," *History Magazine*, 2010, 39–42.
21. Rebecca Weber, "The Travel Agent Is Dying, but It's Not yet Dead—CNN.com," *CNN*, accessed May 18, 2015, http://www.cnn.com/2013/10/03/travel/travel-agent-survival/index.html.
22. Björn Nykvist and M\a ans Nilsson, "Rapidly Falling Costs of Battery Packs for Electric Vehicles," *Nature Climate Change*, 2015, http://www.nature.com/nclimate/journal/vaop/ncurrent/full/nclimate2564.html.
23. Smith, *An Inquiry into the Nature and Causes of the Wealth of Nations* (A. and C. Black, 1863).
24. There are about 7 acres of land per person in the United States, so there is more than enough land for everyone.
25. K. Eric Drexler, *Nanosystems: Molecular Machinery, Manufacturing, and Computation* (New York: John Wiley, 1992).
26. Ray Kurzweil, *The Singularity Is Near* (New York: Penguin Books, 2005).
27. B. T. Wittbrodt et al., "Life-Cycle Economic Analysis of Distributed Manufacturing with Open-Source 3-D Printers," *Mechatronics* 23, 6 (2013): 713–726.
28. Ibid.
29. Kira, "Exclusive: WinSun China Builds World's First 3D Printed Villa and Tallest 3D Printed Apartment Building," *www.3ders.org*, January 18, 2015, http://www.3ders.org/articles/20150118-win-sun-builds-world-first-3d-printed-villa-and-tallest-3d-printed-building-in-china.html.
30. Megan Gannon, News Gannor, "3D Printer Could Transform Moon Dirt Into Lunar Base," *Space.com*, accessed May 18, 2015, http://www.space.com/18694-moon-dirt-3d-printing-lunar-base.html.

31. "The owl of Minerva takes its flight only when the shades of night are gathering," Georg Wilhelm Friedrich Hegel, *Philosophy of Right* (London: George Bell and Sons, 1896).
32. "USDA Economic Research Service—Detail," accessed May 18, 2015, http://www.ers.usda.gov/data-products/chart-gallery/detail.aspx?chartId=40094&ref=collection&embed=True.
33. Drexler, *Nanosystems.*
34. Andy Greenberg, "How 3-D Printed Guns Evolved Into Serious Weapons in Just One Year," *WIRED*, May 15, 2014, http://www.wired.com/2014/05/3d-printed-guns/.

9 Concluding Unscientific Postscript

1. For some interesting thoughts on the political feasibility in the UK of a BIG at present, see Donald Hirsch, "Could a 'Citizen's Income' Work?" (Joseph Rowntree Foundation, 2015), http://www.jrf.org.uk/sites/files/jrf/citizens-income-full.pdf.
2. For more on the Alaska permanent fund see, Karl Widerquist and Michael W. Howard, *Alaska's Permanent Fund Dividend: Examining Its Suitability as a Model* (New York: Palgrave Macmillan, 2012). For thoughts on the value of a guaranteed income less than necessary for financial independence, see Jason B. Murphy, "Baby Steps: Basic Income and the Need for Incremental Organizational Development," *Basic Income Studies* 5, 1 (2010).
3. I borrow these two examples from Karl Widerquist. Karl Widerquist, "OPINION: Big Changes Come," *BIEN*, June 17, 2013, http://www.basicincome.org/news/2013/06/opinion-big-changes-come/.
4. Jurgen De Wispelaere and A. Noguera, "On the Political Feasibility of Universal Basic Income: An Analytic Framework," *Basic Income Guarantee and Politics: International Experiences and Perspectives on the Viability of Income Guarantee* (2012), 17. Jurgen De Wispelaere, "The Struggle for Strategy: On the Politics of Universal Basic Income," *Politics* (2013), http://works.bepress.com/dewispelaere/38/.

References

"2012 HHS Poverty Guidelines." *U.S. Department of Health and Human Services*. Accessed May 10, 2012. http://aspe.hhs.gov/poverty/12poverty.shtml.

"2014 Land Report 100." The Land Report: The Magazine of the American Landowner, n.d. http://www.landreport.com/americas-100-largest-landowners/.

Ackerman, Bruce, and Anne Alstott. *The Stakeholder Society*. New Haven, CT: Yale University Press, 1999.

Anderson, Elizabeth S. "What Is the Point of Equality?" *Ethics* 109, 2 (1999): 287–337.

"Annual Inflation Adjustments for 2013." *Internal Revenue Service*, January 11, 2013. http://www.irs.gov/uac/Newsroom/Annual-Inflation-Adjustments-for-2013.

Bacevich, Andrew J. *The New American Militarism: How Americans Are Seduced by War*. New York: Oxford University Press, 2013.

Bakshi, Gurdip S., and Zhiwu Chen. "The Spirit of Capitalism and Stock-Market Prices." *The American Economic Review*, 1996, 133–157.

Becker, Lawrence C. "The Obligation to Work." *Ethics*, 91, 11980, 35–49.

Berlin, Isaiah. *Four Essays on Liberty*. Oxford: Oxford University Press, 1969.

———. "Two Concepts of Liberty." In *Four Essays on Liberty*, 118–172. Oxford: Oxford University Press, 1969.

Birnbaum, Simon. *Basic Income Reconsidered: Social Justice, Liberalism, and the Demands of Equality*. New York: Palgrave Macmillan, 2012.

Blodget, Henry. "CEO of Apple Partner Foxconn: 'Managing One Million Animals Gives Me A Headache.'" *Business Insider*, 2012. http://www.businessinsider.com/foxconn-animals-2012-1.

Blume, Lawrence, and Steven Durlauf. *The New Palgrave Dictionary of Economics*. New York: Palgrave Macmillan, 2008.

Bolton, Patrick, Joseph F. Brodley, and Michael H. Riordan. "Predatory Pricing: Strategic Theory and Legal Policy." *Georgetown Law Journal* 88 (2000): 2239–2330.

Bronstein, Jamie. "A History of the BIG Idea: Winstanley, Paine, Skidmore and Bellamy." *Journal of Evolution & Technology* 24, 1 (2014). http://jetpress.org/v24/bronstein.htm.

Brynjolfsson, Erik, and Andrew McAfee. *Race against the Machine: How the Digital Revolution Is Accelerating Innovation, Driving Productivity, and*

Irreversibly Transforming Employment and the Economy. Lexington, MA: Digital Frontier Press, 2011.

Bureau of Labor Statistics. "Labor Force Statistics from the Current Population Survey." *Department of Labor*, 2010. http://www.bls.gov/cps/cps_htgm.htm.

"Capitalism." *Merriam-Webster*. Accessed May 12, 2015. http://www.merriam-webster.com/dictionary/capitalism.

Carter, Ian. "Positive and Negative Liberty," *The Stanford Encyclopedia of Philosophy*, ed. Edward N. Zalta, 2003. http://stanford.library.usyd.edu.au/entries/liberty-positive-negative/.

Chan, Casey. "Sushi Restaurant Uses Sushi Robots and Control Centers to Cut Costs." *Gizmodo*, 2010. http://gizmodo.com/5722188/sushi-restaurant-uses-sushi-robots-and-control-centers-to-cut-costs.

Chang, Clio. "Give Families With Children Cash to Help Them Out of Poverty—US News." *US News & World Report*, November 5, 2014. http://www.usnews.com/opinion/economic-intelligence/2014/11/05/give-families-with-children-cash-to-help-them-out-of-poverty.

Christman, J. "Autonomy." *Stanford Encyclopedia of Philosophy*,2 009.

Clark, Gregory. *A Farewell to Alms: A Brief Economic History of the World*. Princeton, NJ: Princeton University Press, 2008.

Cohen, Gerald Allan. *Self-Ownership, Freedom, and Equality*. New York: Cambridge University Press, 1995.

———. *Why Not Socialism?*. Princeton, NJ: Princeton University Press, 2009.

Collender, Stan. "Stop And Smell The Roses: Final 2014 Federal Deficit Fell…Big Time." *Forbes*, October 16, 2014. http://www.forbes.com/sites/stancollender/2014/10/16/stop-and-smell-the-roses-final-2014-federal-deficit-fell-big-time/.

Corneo, Giacomo. "Work Norms and the Welfare State." *CESifo Economic Studies* 58, 4 (December 1, 2012): 599–625. doi:10.1093/cesifo/ifs017.

"Detailed National accounts/GDP Statistics for More than 100 Countries—Charts, Time Series, Historical Data." *Bluenomics*. Accessed May 11, 2015. https://www.bluenomics.com/data#!data/national_accounts_gdp/gdp_expenditure_approach/structure_of_gdp_expenditure_approach/structure_of_gdp_expenditure_approach_annual_nominal_of_gdp|chart/line&countries=usa&sorting=list//title.

De Wispelaere, Jurgen. "Sharing Job Resources: Ethical Reflections on the Justification of Basic Income." *Analyse & Kritik* 22, 2 (2000): 237–256.

———. "The Struggle for Strategy: On the Politics of Universal Basic Income." *Politics*, 2013. http://works.bepress.com/dewispelaere/38/.

De Wispelaere, Jurgen, and A. Noguera. "On the Political Feasibility of Universal Basic Income: An Analytic Framework." In *Basic Income Guarantee and Politics: International Experiences and Perspectives on the Viability of Income Guarantee*, ed. Richard K Caputo, 17–38. New York: Palgrave Macmillan, 2012.

Diener, E. "Subjective Well-Being: The Science of Happiness and a Proposal for a National Index." *American Psychologist* 55, 1 (2000): 34.

Diener, E., R. A. Emmons, R. J. Larsen, and S. Griffin. "The Satisfaction with Life Scale." *Journal of Personality Assessment* 49, 1 (1985): 71–75.

Diener, E., C. Nickerson, R. E. Lucas, and E. Sandvik. "Dispositional Affect and Job Outcomes." *Social Indicators Research* 59, 3 (2002): 229–259.

Domènech, Antoni, and Daniel Raventós. "Property and Republican Freedom: An Institutional Approach to Basic Income." *Basic Income Studies* 2, 2 (2007).

Drexler, K. Eric. *Nanosystems: Molecular Machinery, Manufacturing, and Computation*. New York: John Wiley, 1992.

Easterlin, R. A. *Does Economic Growth Improve the Human Lot? Nations and Households in Economic Growth: Essays in Honour of Moses Abramovitz. M. Abramovitz, PA David and MW Reder*. New York; London: Academic Press, 1974.

Epstein, Richard A. "Against Redress." *Daedalus*, 2002, 39–48.

Feinberg, Joel. "A Child's Right to an Open Future." In *Whose Child? Children's Rights, Parental Authority and State Power*, ed. William Aiken and Hugh LaFollette, 124–153. New Jersey: Rowman & Littlefield, 1980.

Feldman, F. *What Is This Thing Called Happiness?*. New York: Oxford University Press, 2010.

Frankfurt, Harry G. *Freedom of the Will and the Concept of a Person*. Clifton, NJ: Springer, 1988.

Frey, Bruno S. *Happiness: A Revolution in Economics*. Cambridge, MA: MIT Press, 2008.

Frey, Carl Benedikt, and Michael A. Osborne. "The Future of Employment: How Susceptible Are Jobs to Computerisation?" *Retrieved September*, 2013. http://assets.careerspot.com.au/files/news/The_Future_of_Employment_OMS_Working_Paper_1.pdf.

Friedman, Milton. *Capitalism and Freedom*. Chicago: University of Chicago Press, 1962.

Friedman, Milton, and Rose Friedman. *Free to Choose: A Personal Statement*. Harcourt: Houghton Mifflin, 1990.

Gannor, Megan Gannon. News, "3D Printer Could Transform Moon Dirt Into Lunar Base." *Space.com*. Accessed May 18, 2015. http://www.space.com/18694-moon-dirt-3d-printing-lunar-base.html.

Gaus, Gerald F. "Justificatory Liberalism: An Essay on Epistemology and Political Theory." New York: Oxford University Press 1996.

———. *Political Concepts and Political Theories*. Boulder, CO: Westview Press, 2000.

Gonzales, Ralph, Daniel C. Malone, Judith H. Maselli, and Merle A. Sande. "Excessive Antibiotic Use for Acute Respiratory Infections in the United States." *Clinical Infectious Diseases* 33, 6 (2001): 757–762.

Goodin, Robert E. *Utilitarianism as a Public Philosophy*. New York: Cambridge University Press, 1995.

Graham, Carol. *Happiness around the World: The Paradox of Happy Peasants and Miserable Millionaires*. New York: Oxford University Press, 2009.

Green, Thomas Hill. *Lectures on the Principles of Political Obligation and Other Writings*. New York: Cambridge University Press, 1986.

Greenberg, Andy. "How 3-D Printed Guns Evolved Into Serious Weapons in Just One Year." *WIRED*, May 15, 2014. http://www.wired.com/2014/05/3d-printed-guns/.

Growing a Nation. "Historical Time Line," 2005. http://www.agclassroom.org/gan/timeline/.

Gruber, Jon, and Emmanuel Saez. "The Elasticity of Taxable Income: Evidence and Implications." *Journal of Public Economics* 84, 1 (2002): 1–32.

Haybron, D. "Happiness and Pleasure." *Philosophy and Phenomenological Research* 62, 3 (2001): 501–528.

———. "On Being Happy or Unhappy." *Philosophy and Phenomenological Research* 71, 2 (2005): 287–317.

———. *The Pursuit of Unhappiness*. New York: Oxford University Press, 2008.

Hayek, Friedrich A. *The Constitution of Liberty*. Chicago, IL: Chicago University Press, 1960.

"Health as a Share of GDP, 2000–2011, Selected G7 Countries." *OECD Library*, October 31, 2013. http://www.keepeek.com/Digital-Asset-Management/oecd/social-issues-migration-health/health-at-a-glance-2013/health-expenditure-as-a-share-of-gdp-2000-11-selected-g7-countries_health_glance-2013-graph153-en#page1.

Hegel, Georg Wilhelm Friedrich. *Philosophy of Right*. London: George Bell and Sons, 1896.

Hind, Andrew. "Brrrr! History of the Ice Industry." *History Magazine*, 2010, 39–42.

Hirsch, Donald. "Could a 'Citizen's Income' Work?." Joseph Rowntree Foundation, 2015. http://www.jrf.org.uk/sites/files/jrf/citizens-income-full.pdf.

Hobbes, Thomas. *Leviathan: On the Matter, Forme and Power of a Commonwealth and Civil*. New York: Collier, 1962.

"Hoosier Lottery." Accessed May 17, 2015. https://www.hoosierlottery.com/.

Huang, Chye-Ching. "Recent Studies Find Raising Taxes on High-Income Households Would Not Harm the Economy." *Center on Budget and Policy Priorities*, 2012. http://www.cbpp.org/cms/index.cfm?fa=view&id=3756.

Hurka, Thomas. *Perfectionism*. New York: Oxford University Press, 1993.

———. "Review: [untitled]." *Mind*, New Series, 110, 439 (July 2001): 878–881.

Huxley, A. *Brave New World and Brave New World Revisited.* Toronto, Canada: Vintage Books, 2007.

Iacocca, Lee A., Sonny Kleinfield, and Ray McKenzie. *Talking Straight.* New York: Bantam books, 1988.

Kahneman, Daniel, and Angus Deaton. "High Income Improves Evaluation of Life but Not Emotional Well-Being." *Proceedings of the National Academy of Sciences* 107, 38 (2010): 16489–93.

Keller, Simon. "Welfarism." *Philosophy Compass* 4, 1 (2009): 82–95.

Kelly, Kevin. "Better Than Human: Why Robots Will—And Must—Take Our Jobs." *Wired*, 2012. http://www.wired.com/gadgetlab/2012/12/ff-robots-will-take-our-jobs/all/.

Kershnar, Stephen. "A Liberal Argument for Slavery." *Journal of Social Philosophy* 34, 4 (2003): 510–536.

"Key Issues: Tax Expenditures." *Government Accountability Office.* Accessed May 11, 2015. http://www.gao.gov/key_issues/tax_expenditures/issue_summary.

Khosla, Vinod. "Technology Will Replace 80% of What Doctors Do." *CNN*, 2102. http://tech.fortune.cnn.com/2012/12/04/technology-doctors-khosla/.

Kira. "Exclusive: WinSun China Builds World's First 3D Printed Villa and Tallest 3D Printed Apartment Building." *www.3ders.org*, January 18, 2015. http://www.3ders.org/articles/20150118-winsun-builds-world-first-3d-printed-villa-and-tallest-3d-printed-building-in-china.html.

Krueger, Alan B., and David A. Schkade. "The Reliability of Subjective Well-Being Measures." *Journal of Public Economics* 92, 8 (2008): 1833–45.

Krugman,Paul."DegreesandDollars." *NewYorkTimes*,March6,2011.http://www.lib.tku.edu.tw/service/exampdf/dayexam/01day/10143027.pdf.

Kuflik, Arthur. "The Inalienability of Autonomy." *Philosophy & Public Affairs*, 1984, 271–298.

Kurzweil, Ray. *The Singularity Is Near.* New York: Penguin Books, 2005.

Kymlicka, Will. *Contemporary Political Philosophy.* Oxford: Oxford University Press, 2001.

Lazear, Edward P., and James R. Spletzer. *The United States Labor Market: Status Quo or a New Normal?*. National Bureau of Economic Research, 2012.

Lee, Timorthy. "Amazon Envisions Eventually Delivering Packages in 30 Minutes via Drones." *The Washington Post*, 2013. http://www.washingtonpost.com/blogs/the-switch/wp/2013/12/01/amazon-wants-to-deliver-packages-in-30-minutes-with-drones/.

Leonhardt, David. "Maybe Money Does Buy Happiness After All." *The New York Times*, April 16, 2008, sec. Business. http://www.nytimes.com/2008/04/16/business/16leonhardt.html.

Levy, David. "The Ethics of Robot Prostitutes." In *Robot Ethics: The Ethical and Social Implications of Robotics*, ed. Patrick Lin, Keith Abney, and George A Bekey, 223–232. Cambridge, MA: MIT Press, 2011. Liptak,

Adam. "1 in 100 US Adults behind Bars, New Study Says." *New York Times*, December 9, 28, 2008. http://sites.google.com/a/trammellcriminology.com/www/NewYorkTimesArticle.pdf.

"Living Wage Calculation for Las Cruces City, Dona Ana County, New Mexico." *Living Wage Calculator*. Accessed May 10, 2015. http://livingwage.mit.edu/places/3501339380.

Locke, J. *Two Treatises of Government*. New York: Cambridge University Press, 2008.

Lovett, Frank. "Domination and Distributive Justice." *The Journal of Politics* 71, 3 (2009): 817–830.

Lowrey, Annie. "C.B.O. Cuts 2013 Deficit Estimate by 24%." *The New York Times*, May 14, 2013. http://www.nytimes.com/2013/05/15/business/cbo-cuts-2013-deficit-estimate-by-24-percent.html.

Lyubomirsky, S., L. King, and E. Diener. "The Benefits of Frequent Positive Affect: Does Happiness Lead to Success?" *Psychological Bulletin* 131 (2005): 803–855.

Lyubomirsky, Sonja, and Heidi S. Lepper. "A Measure of Subjective Happiness: Preliminary Reliability and Construct Validation." *Social Indicators Research* 46, 2 (1999): 137–155.

Markoff, John. "Skilled Work, Without the Worker." *New York Times*, 2012. http://www.nytimes.com/2012/08/19/business/new-wave-of-adept-robots-is-changing-global-industry.html?pagewanted=all&_r=0.

Marx, Karl, and Friedrich Engels. *The Communist Manifesto*. Peking: Foreign Language Press, 1975.

Mattel, Deborah, and Charles Schafer. "An Investigation of Validity of the Subjective Happiness Scale 1." *Psychological Reports* 94, 1 (2004): 288–290.

McKinnon, Catriona. "Basic Income, Self-Respect and Reciprocity." *Journal of Applied Philosophy* 20, 2 (2003): 143–158.

Milgrom, Paul. "Predatory Pricing." *The New Palgrave Dictionary of Economics* 3 (1987): 938.

Morris, Harvey. "U.S. Health Care Costs More Than 'Socialized' European Medicine." *IHT Rendezvous*. Accessed May 12, 2015. http://rendezvous.blogs.nytimes.com/2012/06/28/u-s-health-care-costs-more-than-socialized-european-medicine/.

Motta, Massimo. *Competition Policy: Theory and Practice*. New York: Cambridge University Press, 2004.

Murphy, Jason B. "Baby Steps: Basic Income and the Need for Incremental Organizational Development." *Basic Income Studies* 5, 1 (2010).

Murray, Peter. "Robot Serves Up 360 Hamburgers Per Hour." *Singularity Hub*, 2013. http://singularityhub.com/2013/01/22/robot-serves-up-340-hamburgers-per-hour/.

Musgrave, Richard A. "Maximin, Uncertainty, and the Leisure Trade-Off." *The Quarterly Journal of Economics* 88, 4 (1974): 625–632.

Nielsen, Kai. *Equality and Liberty: A Defense of Radical Egalitarianism.* Totowa, NJ: Rowman & Littlefield, 1984.

Nolte, Ellen, and C. Martin McKee. "In Amenable Mortality—Deaths Avoidable through Health Care—Progress in the US Lags That of Three European Countries." *Health Affairs*, 2012, doi: 10.1377

Nozick, R. *Anarchy, State and Utopia.* New York, NY: Basic Books, 1974.

———. *Philosophical Explanations.* Cambridge, MA: Belknap Press, 1981.

Nykvist, Björn, and M\a ans Nilsson. "Rapidly Falling Costs of Battery Packs for Electric Vehicles." *Nature Climate Change* (2015). http://www.nature.com/nclimate/journal/vaop/ncurrent/full/nclimate2564.html.

OECD. 39–71. Organisation for Economic Co-operation and Development, 2008. http://www.oecd-ilibrary.org/content/chapter/ctt-2008-4-en.

Oshana, Marina. *Personal Autonomy in Society.* Burlington, VT: Ashgate, 2006.

Paine, Thomas. *Agrarian Justice.* Raleigh, NC: W. Adlard, 1974.

Parker, Ashley. "House Republicans Cheer Boehner's Refusal to Negotiate on Cuts." *The New York Times*, February 28, 2013. http://www.nytimes.com/2013/03/01/us/politics/house-republicans-cheer-boehners-refusal-to-negotiate-on-cuts.html.

Paul, Ron. "Imagine." *From Reason to Freedom*, March 9, 2009. http://www.reasontofreedom.com/node/1775.

"Per Capita Income by State." *Bureau of Business and Economic Research, UNM*, April 2, 2013. https://bber.unm.edu/econ/us-pci.htm.

Perez, Bien. "Hon Hai Plan to Deploy 1m Robots Faces Delay, Says Analyst." *South China Morning Post*, 2012. http://www.scmp.com/business/companies/article/1052557/hon-hai-plan-deploy-1m-robots-faces-delay-says-analyst.

Perlo-Freeman, Sam, Elisabeth Sköns, Carina Solmirano, and Helén Wilandh. *Trends in World Military Expenditure, 2012.* Solna, Sweden: Stockholm International Peace Research Institute, 2013.

Pettit, Philip. "A Republican Right to Basic Income?" *Basic Income Studies* 2, 2 (2007): 1–8.

Piketty, Thomas. *Capital in the Twenty-First Century.* Cambridge, MA: Harvard University Press, 2014.

"Population by Sex, Rate of Population Increase, Surface Area and Density." *United Nations Stats*, 2006. http://unstats.un.org/unsd/demographic/products/dyb/dyb2006/Table03.pdf.

Rawls, J. *A Theory of Justice.* Cambridge, MA: Harvard University Press, 1971.

Rawls, John. *Justice as Fairness: A Restatement.* Cambridge, MA: Harvard University Press, 2001.

———. *Political Liberalism: Expanded Edition.* New York: Columbia University Press, 2005.

———. "The Priority of Right and Ideas of the Good." *Philosophy & Public Affairs* 17, 4 (1988): 251–276.

Raz, Joseph. *The Morality of Freedom*. New York: Oxford University Press, 1986.

Reiman, J. *As Free and as Just as Possible: The Theory of Marxian Liberalism*. Malden, MA: Wiley-Blackwell, 2012.

Reiman, Jeffrey, and Paul Leighton. *The Rich Get Richer and the Poor Get Prison*. 10th ed. New York: Pearson, 2012.

Rockoff, J. D. "Robots vs. Anesthesiologists." *Wall Street Journal*, Pg. B8–9, 2013.

Roediger, David R. and Philip Sheldon Foner. *Our Own Time: A History of American Labor and the Working Day*. London, UK: Verso, 1989.

Rousseau, Jean-Jacques. *The Social Contract*. Great Britain: Penguin, 1971.

Roy, Raja. "Exploring the Boundary Conditions of Disruption: Large Firms and New Product Introduction With a Potentially Disruptive Technology in the Industrial Robotics Industry," 2014. http://ieeexplore.ieee.org/xpls/abs_all.jsp?arnumber=6578147.

Rucker, Philip. "GOP Senate Hopeful in Alaska Preaches 'more Freedom, Less Government.'" *The Washington Post*, October 10, 2014. http://www.washingtonpost.com/politics/senate-hopefuls-say-theyll-show-washington-the-alaskan-way/2014/10/10/7891dd6a-4e97-11e4-babe-e91da079cb8a_story.html.

"Rudy Turns Up the Heat—The Rush Limbaugh Show." *Rush Limbaugh*, April 25, 2007. http://www.rushlimbaugh.com/daily/2007/04/25/rudy_turns_up_the_heat.

Russell, B. "We Don't Want to Be Happy." In *Aldous Huxley*, ed. C. Watt. London, UK: Psychology Press, 1997.

Russell, Bertrand. *Roads to Freedom: Socialism, Anarchism and Syndicalism*. London, UK: Allen & Unwin, 1966.

Rytina, Nancy. *Homeland Security Office of Immigration Statistics*, n.d. http://www.dhs.gov/sites/default/files/publications/ois_lpr_pe_2012.pdf.

Sacks, Daniel W., Betsey Stevenson, and Justin Wolfers. "The New Stylized Facts about Income and Subjective Well-Being." *Emotion* 12, 6 (2012): 1181–1187.

Sager, Ira. "Farm Bots and Two-Month-Old Bread: Innovations in Food and Agriculture," 2013. http://www.businessweek.com/articles/2013-01-10/farm-bots-and-two-month-old-bread-innovations-in-food-and-agriculture.

Schiller, Bradley R., and others. *The Economics of Poverty and Discrimination*. Upper Saddle River, NJ: Prentice Hall, 2001.

Schutz, Robert R. *The $30,000 Solution: A Guaranteed Annual Income for Every American*. Santa Barbara, CA: Fithian Press, 1996.

Sheahen, A. *It's Time to Think BIG! How to Simplify the Tax Code and Provide Every American With a Basic Income Guarantee*. USBIG Discussion Paper, 2006. http://usbig.net/papers/144-Sheahen-RefundableTax-Credit.pdf.

Sheahen, Allan. *Basic Income Guarantee: Your Right to Economic Security.* New York: Palgrave Macmillan, 2012.

Smith, Adam. *An Inquiry into the Nature and Causes of the Wealth of Nations.* Edinburgh, Scottland: A. and C. Black, 1863.

Smith, Eric Alden, Monique Borgerhoff Mulder, Samuel Bowles, Michael Gurven, Tom Hertz, and Mary K. Shenk. "Production Systems, Inheritance, and Inequality in Premodern Societies." *Current Anthropology* 51, 1 (2010): 85–94.

Sofge, Erik. "3 New Farm Bots Programmed to Pick, Plant and Drive." *Popular Mechanics*, 2009. http://www.popularmechanics.com/technology/engineering/robots/4328685.

Standing, G. *The Precariat: The New Dangerous Class.* New York: Bloomsbury, 2011.

"State & County Quick Facts." *The United States Census Bureau*, April 29, 2015. http://quickfacts.census.gov/qfd/states/00000.html.

Stevenson, Betsey, and Justin Wolfers. *Subjective Well-Being and Income: Is There Any Evidence of Satiation?*. National Bureau of Economic Research, 2013. http://www.nber.org/papers/w18992.

Sumner, L. W. "Two Theories of the Good." In *The Good Life and the Human Good*, ed. Ellen Frankel Paul, Fred D. Miller Jr., and Jeffrey Paul, 1–15. New York: Cambridge University Press, 1992.

———. *Welfare, Happiness, and Ethics.* New York: Oxford University Press, 1996.

Sumner, Wayne. *The Moral Foundation of Rights.* New York: Oxford University Press, 1987.

Suplicy, Eduardo M., and Christovam Buarque. "A Guaranteed Minimum Income to Eradicate Poverty and Help Poor Children Go to School instead of Being Forced to Work. The Brazilian Debate and Experience." In *BIEN's 6th Congress,* paper presented at BIEN'S 6th Congress, September12–14, 1996. Vienna, 1996.

Tabarrok, Alex. "Productivity and Unemployment." *Marginal Revolution*, 2003. http://marginalrevolution.com/marginalrevolution/2003/12/productivity_an.html.

Tatarkiewicz, W. *Analysis of Happiness.* Vol. 3. Poland, Warszawa: Kluwer Academic Pub, 1976.

Tawney, R. H. "*Equality (1931)*." London: Allen & Unwin, 1952.

Telfer, E. *Happiness.* New York: St. Martin's Press, 1980.

"The Distribution of Household Income and Federal Taxes, 2010." *Congressional Budget Office.* Accessed May 17, 2015. http://www.cbo.gov/publication/44604.

The New York Times. "Effective Income Tax Rates," 2012. http://www.nytimes.com/interactive/2012/01/18/us/effective-income-tax-rates.html?_r=2&.

Thompson, Jeffrey. *Raising Revenue from High-Income Households: Should States Continue to Place the Lowest Tax Rates on Those with the*

Highest Incomes?. Political Economy Research Institute, University of Massachusetts at Amherst, 2012. https://ideas.repec.org/p/uma/perips/revenue_peri_march5.html.

Trabandt, Mathias, and Harald Uhlig. *How Do Laffer Curves Differ across Countries?*. National Bureau of Economic Research, 2012. http://www.nber.org/papers/w17862.

———. "The Laffer Curve Revisited." *Journal of Monetary Economics* 58, 4 (2011): 305–327.

Turgot, Anne Robert Jacques, and Peter Diderik Groenewegen. *The Economics of ARJ Turgot*. The Hague, Netherlands: Martinus Nijhoff, 1977.

US Census Bureau. "Current Population Survey." *US Census Bureau*, January 14, 2013. http://www.census.gov/hhes/www/cpstables/032011/perinc/new01_001.htm.

"USDA Economic Research Service – Detail." Accessed May 18, 2015. http://www.ers.usda.gov/data-products/chart-gallery/detail.aspx?chartId=40094&ref=collection&embed=True.

US Department of Commerce, B. E. A. Accessed May 11, 2015. http://www.bea.gov/newsreleases/national/gdp/gdpnewsrelease.htm.

Vallentyne, Peter. "Libertarianism and the Justice of a Basic Income." *Basic Income Studies* 6, 2 (2011).

Van Donselaar, Gigs. *The Right to Exploit: Parasitism, Scarcity, and Basic Income*. New York: Oxford University Press, 2009.

Van Donselaar, Gijs. "The Freedom-Based Account of Solidarity and Basic Income." *Ethical Theory and Moral Practice* 1, 3 (1998): 313–333.

Van Parijs, Philippe. "Why Surfers Should Be Fed: The Liberal Case for an Unconditional Basic Income." *Philosophy & Public Affairs* 20, 2 (1991): 101–131.

Van Parijs, Philippe. "Real Freedom for All: What (if Anything) Can Justify Capitalism?" *OUP Catalogue*, 1997.

Veenhoven, R. "Happiness in Nations." *Subjective Appreciation of Life in* 56 (1993): 1946–1992.

———. "Is Happiness Relative?" *Social Indicators Research* 24, 1 (1991): 1–34.

Wagstaff, Keith. "How Foxconn's Million-Machine 'Robot Kingdom' Will Change the Face of Manufacturing." *Time*, 2011. http://techland.time.com/2011/11/09/how-foxconns-million-machine-robot-kingdom-will-change-the-face-of-manufacturing/.

Walker, M. "A Moral Paradox in the Creation of Artificial Intelligence: Mary Poppins 3000s of the World Unite!" *AAAI Press* Human Implications of Human-Robot Interaction: Papers from the AAAI Workshop (2006): 23–28.

———. "BIG and Technological Unemployment: Chicken Little Versus the Economists." *Journal of Evolution & Technology* 24, 1 (2014): 5–25.

———. *Happy-People-Pills for All*. Hoboken, NJ: John Wiley, 2013.

Wall, Steven. *Liberalism, Perfectionism and Restraint*. New York: Cambridge University Press, 1998.

Wang, Ting, Ling Wang, and Tiffany Chen. "Secrets of Success: Innovation in the Silicon Valley: Business Development and Innovation of Tech Companies." In *Behavior, Economic and Social Computing (BESC), 2014 International Conference, Oct. 30–Nov. 1. Shanghai*, 1–5. IEEE, 2014. http://ieeexplore.ieee.org/xpls/abs_all.jsp?arnumber=7059530.

Weber, Rebecca. "The Travel Agent Is Dying, but It's Not yet Dead—CNN.com." *CNN*. Accessed May 18, 2015. http://www.cnn.com/2013/10/03/travel/travel-agent-survival/index.html.

White, Stuart. "Social Minimum," 2004. http://stanford.library.usyd.edu.au/entries/social-minimum/.

———. *The Civic Minimum: On the Rights and Obligations of Economic Citizenship*. New York: Oxford University Press, 2003.

Widerquist, Karl. "A Failure to Communicate: What (if Anything) Can We Learn from the Negative Income Tax Experiments?" *The Journal of Socio-Economics* 34, 1 (2005): 49–81.

———. *Independence, Propertylessness, and Basic Income: A Theory of Freedom as the Power to Say No*. New York: Palgrave Macmillan, 2013.

———. "OPINION: Big Changes Come." *BIEN*, June 17, 2013. http://www.basicincome.org/news/2013/06/opinion-big-changes-come/.

———. "Perspectives on the Guaranteed Income, Part I." *Journal of Economic Issues* (2001): 749–757.

———. "The Physical Basis of Voluntary Trade." *Human Rights Review* 11, 1 (2010): 83–103.

Widerquist, Karl, and Michael W. Howard. *Alaska's Permanent Fund Dividend: Examining Its Suitability as a Model*. New York: Palgrave Macmillan, 2012.

Widerquist, Karl, and Michael A. Lewis. "An Efficiency Argument for the Basic Income Guarantee." *International Journal of Environment, Workplace and Employment* 2, 1 (2006): 21–43.

Williamson, Thad. "Who Owns What? An Egalitarian Interpretation of John Rawls's Idea of a Property-Owning Democracy." *Journal of Social Philosophy* 40, 3 (2009): 434–453.

Wilper, Andrew P., Steffie Woolhandler, Karen E. Lasser, Danny McCormick, David H. Bor, and David U. Himmelstein. "Health Insurance and Mortality in US Adults." *American Journal of Public Health* 99, 12 (2009).

Wittbrodt, B. T., A. G. Glover, J. Laureto, G. C. Anzalone, D. Oppliger, J. L. Irwin, and J. M. Pearce. "Life-Cycle Economic Analysis of Distributed Manufacturing with Open-Source 3-D Printers." *Mechatronics* 23, 6 (2013): 713–726.

Wright, Erik Olin. *Envisioning Real Utopias*. London, UK: Verso, 2010.

Index